Packaging Design

STEM Road Map for Middle School

Packaging Design

STEM Road Map for Middle School

Edited by Carla C. Johnson,
Janet B. Walton, and Erin Peters-Burton

Arlington, Virginia

Claire Reinburg, Director
Rachel Ledbetter, Managing Editor
Deborah Siegel, Associate Editor
Andrea Silen, Associate Editor
Donna Yudkin, Book Acquisitions Manager

Art and Design
Will Thomas Jr., Director, cover and interior design
Himabindu Bichali, Graphic Designer, interior design

Printing and Production
Catherine Lorrain, Director

National Science Teachers Association
David L. Evans, Executive Director

1840 Wilson Blvd., Arlington, VA 22201
www.nsta.org/store
For customer service inquiries, please call 800-277-5300.

21 20 19 18 4 3 2 1

NSTA is committed to publishing material that promotes the best in inquiry-based science education. However, conditions of actual use may vary, and the safety procedures and practices described in this book are intended to serve only as a guide. Additional precautionary measures may be required. NSTA and the authors do not warrant or represent that the procedures and practices in this book meet any safety code or standard of federal, state, or local regulations. NSTA and the authors disclaim any liability for personal injury or damage to property arising out of or relating to the use of this book, including any of the recommendations, instructions, or materials contained therein.

Library of Congress Cataloging-in-Publication Data
Names: Johnson, Carla C., 1969- editor. | Walton, Janet B., 1968- editor. | Peters-Burton, Erin E., editor.
Title: Packaging design, grade 6 : STEM road map for middle school / edited by Carla C. Johnson, Janet B. Walton, and Erin Peters-Burton.
Description: Arlington, VA : National Science Teachers Association, [2018] | Includes bibliographical references and index.
Identifiers: LCCN 2018008716 (print) | LCCN 2018012599 (ebook) | ISBN 9781681404530 (e-book) | ISBN 9781681404523 (print)
Subjects: LCSH: Packaging--Study and teaching (Elementary)--Activity programs | English language--Study and teaching (Elementary)--Activity programs. | Sixth grade (Education)
Classification: LCC TS195 (ebook) | LCC TS195 .P3198 2018 (print) | DDC 658.5/64--dc23
LC record available at *https://lccn.loc.gov/2018008716*

The *Next Generation Science Standards* ("NGSS") were developed by twenty-six states, in collaboration with the National Research Council, the National Science Teachers Association, and the American Association for the Advancement of Science in a process managed by Achieve, Inc. For more information go to *www.nextgenscience.org.*

CONTENTS

CONTENTS

ABOUT THE EDITORS AND AUTHORS

Dr. Carla C. Johnson is the associate dean for research, engagement, and global partnerships and a professor of science education at Purdue University's College of Education in West Lafayette, Indiana. Dr. Johnson serves as the director of research and evaluation for the Department of Defense–funded Army Educational Outreach Program (AEOP), a global portfolio of STEM education programs, competitions, and apprenticeships. She has been a leader in STEM education for the past decade, serving as the director of STEM Centers, editor of the *School Science and Mathematics* journal, and lead researcher for the evaluation of Tennessee's Race to the Top–funded STEM portfolio. Dr. Johnson has published over 100 articles, books, book chapters, and curriculum books focused on STEM education. She is a former science and social studies teacher and was the recipient of the 2013 Outstanding Science Teacher Educator of the Year award from the Association for Science Teacher Education (ASTE), the 2012 Award for Excellence in Integrating Science and Mathematics from the School Science and Mathematics Association (SSMA), the 2014 award for best paper on Implications of Research for Educational Practice from ASTE, and the 2006 Outstanding Early Career Scholar Award from SSMA. Her research focuses on STEM education policy implementation, effective science teaching, and integrated STEM approaches.

Dr. Janet B. Walton is a research assistant professor and the assistant director of evaluation for AEOP at Purdue University's College of Education. Formerly the STEM workforce program manager for Virginia's Region 2000 and founding director of the Future Focus Foundation, a nonprofit organization dedicated to enhancing the quality of STEM education in the region, she merges her economic development and education backgrounds to develop K–12 curricular materials that integrate real-life issues with sound cross-curricular content. Her research focuses on collaboration between schools and community stakeholders for STEM education and problem- and project-based learning pedagogies. With this research agenda, she works to forge productive relationships between K–12 schools and local business and community stakeholders to bring contextual STEM experiences into the classroom and provide students and educators with innovative resources and curricular materials.

ABOUT THE EDITORS AND AUTHORS

Dr. Erin Peters-Burton is the Donna R. and David E. Sterling endowed professor in science education at George Mason University in Fairfax, Virginia. She uses her experiences from 15 years as an engineer and secondary science, engineering, and mathematics teacher to develop research projects that directly inform classroom practice in science and engineering. Her research agenda is based on the idea that all students should build self-awareness of how they learn science and engineering. She works to help students see themselves as "science-minded" and help teachers create classrooms that support student skills to develop scientific knowledge. To accomplish this, she pursues research projects that investigate ways that students and teachers can use self-regulated learning theory in science and engineering, as well as how inclusive STEM schools can help students succeed. During her tenure as a secondary teacher, she had a National Board Certification in Early Adolescent Science and was an Albert Einstein Distinguished Educator Fellow for NASA. As a researcher, Dr. Peters-Burton has published over 100 articles, books, book chapters, and curriculum books focused on STEM education and educational psychology. She received the Outstanding Science Teacher Educator of the Year award from ASTE in 2016 and a Teacher of Distinction Award and a Scholarly Achievement Award from George Mason University in 2012, and in 2010 she was named University Science Educator of the Year by the Virginia Association of Science Teachers.

Toni A. Ivey is an associate professor of science education in the College of Education at Oklahoma State University. A former science teacher, Dr. Ivey's research is focused on science and STEM education for students and teachers across K–20.

Tamara J. Moore is an associate professor of engineering education in the College of Engineering at Purdue University. Dr. Moore's research focuses on defining STEM integration through the use of engineering as the connection and investigating its power for student learning.

Sue Christian Parsons is an associate professor and the Jacques Munroe Professor in Reading and Literacy Education at Oklahoma State University. A former English language arts teacher, her research focuses on teacher development and teaching and advocating for diverse learners through literature for children and young adults.

Dr. Adrienne Redmond-Sanogo is an associate professor of mathematics education in the College of Education at Oklahoma State University. Dr. Redmond-Sanogo's research is focused on mathematics and STEM education across K–12 and preservice teacher education.

Dr. Toni A. Sondergeld is an associate professor of assessment, research, and statistics in the School of Education at Drexel University in Philadelphia. Dr. Sondergeld's research concentrates on assessment and evaluation in education, with a focus on K–12 STEM.

Dr. Juliana Utley is an associate professor and the Morsani Chair in Mathematics Education in the College of Education at Oklahoma State University. A former mathematics teacher, Dr. Utley centers her research on mathematics and STEM education across K–12.

ACKNOWLEDGMENTS

This module was developed as a part of the STEM Road Map project (Carla C. Johnson, principal investigator). The Purdue University College of Education, General Motors, and other sources provided funding for this project.

See *www.routledge.com/products/9781138804234* for more information about *STEM Road Map: A framework for integrated STEM education.*

THE STEM ROAD MAP

BACKGROUND, THEORY, AND PRACTICE

1

OVERVIEW OF THE *STEM ROAD MAP CURRICULUM SERIES*

Carla C. Johnson, Erin Peters-Burton, and Tamara J. Moore

The *STEM Road Map Curriculum Series* was conceptualized and developed by a team of STEM educators from across the United States in response to a growing need to infuse real-world learning contexts, delivered through authentic problem-solving pedagogy, into K–12 classrooms. The curriculum series is grounded in integrated STEM, which focuses on the integration of the STEM disciplines—science, technology, engineering, and mathematics—delivered across content areas, incorporating the Framework for 21st Century Learning along with grade-level-appropriate academic standards.

The curriculum series begins in kindergarten, with a five-week instructional sequence that introduces students to the STEM themes and gives them grade-level-appropriate topics and real-world challenges or problems to solve. The series uses project-based and problem-based learning, presenting students with the problem or challenge during the first lesson, and then teaching them science, social studies, English language arts, mathematics, and other content, as they apply what they learn to the challenge or problem at hand.

Authentic assessment and differentiation are embedded throughout the modules. Each *STEM Road Map Curriculum Series* module has a lead discipline, which may be science, social studies, English language arts, or mathematics. All disciplines are integrated into each module, along with ties to engineering. Another key component is the use of STEM Research Notebooks to allow students to track their own learning progress. The modules are designed with a scaffolded approach, with increasingly complex concepts and skills introduced as students progress through grade levels.

The developers of this work view the curriculum as a resource that is intended to be used either as a whole or in part to meet the needs of districts, schools, and teachers who are implementing an integrated STEM approach. A variety of implementation formats are possible, from using one stand-alone module at a given grade level to using all five modules to provide 25 weeks of instruction. Also, within each grade band (K–2, 3–5, 6–8, 9–12), the modules can be sequenced in various ways to suit specific needs.

STANDARDS-BASED APPROACH

The *STEM Road Map Curriculum Series* is anchored in the *Next Generation Science Standards* (*NGSS*), the *Common Core State Standards for Mathematics* (*CCSS Mathematics*), the *Common Core State Standards for English Language Arts* (*CCSS ELA*), and the Framework for 21st Century Learning. Each module includes a detailed curriculum map that incorporates the associated standards from the particular area correlated to lesson plans. The STEM Road Map has very clear and strong connections to these academic standards, and each of the grade-level topics was derived from the mapping of the standards to ensure alignment among topics, challenges or problems, and the required academic standards for students. Therefore, the curriculum series takes a standards-based approach and is designed to provide authentic contexts for application of required knowledge and skills.

THEMES IN THE *STEM ROAD MAP CURRICULUM SERIES*

The K–12 STEM Road Map is organized around five real-world STEM themes that were generated through an examination of the big ideas and challenges for society included in STEM standards and those that are persistent dilemmas for current and future generations:

- Cause and Effect
- Innovation and Progress
- The Represented World
- Sustainable Systems
- Optimizing the Human Experience

These themes are designed as springboards for launching students into an exploration of real-world learning situated within big ideas. Most important, the five STEM Road Map themes serve as a framework for scaffolding STEM learning across the K–12 continuum.

The themes are distributed across the STEM disciplines so that they represent the big ideas in science (Cause and Effect; Sustainable Systems), technology (Innovation and Progress; Optimizing the Human Experience), engineering (Innovation and Progress; Sustainable Systems; Optimizing the Human Experience), and mathematics (The Represented World), as well as concepts and challenges in social studies and 21st century skills that are also excellent contexts for learning in English language arts. The process of developing themes began with the clustering of the *NGSS* performance expectations and the National Academy of Engineering's grand challenges for engineering, which led to the development of the challenge in each module and connections of the module activities to the *CCSS Mathematics* and *CCSS ELA* standards. We performed these mapping processes with large teams of experts and found that these five themes provided breadth, depth, and coherence to frame a high-quality STEM learning experience from kindergarten through 12th grade.

Cause and Effect

The concept of cause and effect is a powerful and pervasive notion in the STEM fields. It is the foundation of understanding how and why things happen as they do. Humans spend considerable effort and resources trying to understand the causes and effects of natural and designed phenomena to gain better control over events and the environment and to be prepared to react appropriately. Equipped with the knowledge of a specific cause-and-effect relationship, we can lead better lives or contribute to the community by altering the cause, leading to a different effect. For example, if a person recognizes that irresponsible energy consumption leads to global climate change, that person can act to remedy his or her contribution to the situation. Although cause and effect is a core idea in the STEM fields, it can actually be difficult to determine. Students should be capable of understanding not only when evidence points to cause and effect but also when evidence points to relationships but not direct causality. The major goal of education is to foster students to be empowered, analytic thinkers, capable of thinking through complex processes to make important decisions. Understanding causality, as well as when it cannot be determined, will help students become better consumers, global citizens, and community members.

Innovation and Progress

One of the most important factors in determining whether humans will have a positive future is innovation. Innovation is the driving force behind progress, which helps create possibilities that did not exist before. Innovation and progress are creative entities, but in the STEM fields, they are anchored by evidence and logic, and they use established concepts to move the STEM fields forward. In creating something new, students must consider what is already known in the STEM fields and apply this knowledge appropriately. When we innovate, we create value that was not there previously and create new conditions and possibilities for even more innovations. Students should consider how their innovations might affect progress and use their STEM thinking to change current human burdens to benefits. For example, if we develop more efficient cars that use by-products from another manufacturing industry, such as food processing, then we have used waste productively and reduced the need for the waste to be hauled away, an indirect benefit of the innovation.

The Represented World

When we communicate about the world we live in, how the world works, and how we can meet the needs of humans, sometimes we can use the actual phenomena to explain a concept. Sometimes, however, the concept is too big, too slow, too small, too fast, or too complex for us to explain using the actual phenomena, and we must use a representation or a model to help communicate the important features. We need representations and models such as graphs, tables, mathematical expressions, and diagrams because it makes our thinking visible. For example, when examining geologic time, we cannot actually observe the passage of such large chunks of time, so we create a timeline or a model that uses a proportional scale to visually illustrate how much time has passed for different eras. Another example may be something too complex for students at a particular grade level, such as explaining the *p* subshell orbitals of electrons to fifth graders. Instead, we use the Bohr model, which more closely represents the orbiting of planets and is accessible to fifth graders.

When we create models, they are helpful because they point out the most important features of a phenomenon. We also create representations of the world with mathematical functions, which help us change parameters to suit the situation. Creating representations of a phenomenon engages students because they are able to identify the important features of that phenomenon and communicate them directly. But because models are estimates of a phenomenon, they leave out some of the details, so it is important for students to evaluate their usefulness as well as their shortcomings.

Sustainable Systems

From an engineering perspective, the term *system* refers to the use of "concepts of component need, component interaction, systems interaction, and feedback. The interaction of subcomponents to produce a functional system is a common lens used by all engineering disciplines for understanding, analysis, and design." (Koehler, Bloom, and Binns 2013, p. 8). Systems can be either open (e.g., an ecosystem) or closed (e.g., a car battery). Ideally, a system should be sustainable, able to maintain equilibrium without much energy from outside the structure. Looking at a garden, we see flowers blooming, weeds sprouting, insects buzzing, and various forms of life living within its boundaries. This is an example of an ecosystem, a collection of living organisms that survive together, functioning as a system. The interaction of the organisms within the system and the influences of the environment (e.g., water, sunlight) can maintain the system for a period of time, thus demonstrating its ability to endure. Sustainability is a desirable feature of a system because it allows for existence of the entity in the long term.

In the STEM Road Map project, we identified different standards that we consider to be oriented toward systems that students should know and understand in the K–12 setting. These include ecosystems, the rock cycle, Earth processes (such as erosion, tectonics, ocean currents, weather phenomena), Earth-Sun-Moon cycles, heat transfer, and the interaction among the geosphere, biosphere, hydrosphere, and atmosphere. Students and teachers should understand that we live in a world of systems that are not independent of each other, but rather are intrinsically linked such that a disruption in one part of a system will have reverberating effects on other parts of the system.

Optimizing the Human Experience

Science, technology, engineering, and mathematics as disciplines have the capacity to continuously improve the ways humans live, interact, and find meaning in the world, thus working to optimize the human experience. This idea has two components: being more suited to our environment and being more fully human. For example, the progression of STEM ideas can help humans create solutions to complex problems, such as improving ways to access water sources, designing energy sources with minimal impact on our environment, developing new ways of communication and expression, and building efficient shelters. STEM ideas can also provide access to the secrets and wonders of nature. Learning in STEM requires students to think logically and systematically, which is a way of knowing the world that is markedly different from knowing the world as an artist. When students can employ various ways of knowing and understand when it is appropriate to use a different way of knowing or integrate ways of knowing, they are fully experiencing the best of what it is to be human. The problem-based learning scenarios provided in the STEM Road Map help students develop ways of thinking like STEM professionals as they ask questions and design solutions. They learn to optimize the human experience by innovating improvements in the designed world in which they live.

THE NEED FOR AN INTEGRATED STEM APPROACH

At a basic level, STEM stands for science, technology, engineering, and mathematics. Over the past decade, however, STEM has evolved to have a much broader scope and implications. Now, educators and policy makers refer to STEM as not only a concentrated area for investing in the future of the United States and other nations but also as a domain and mechanism for educational reform.

The good intentions of the recent decade-plus of focus on accountability and increased testing has resulted in significant decreases not only in instructional time for teaching science and social studies but also in the flexibility of teachers to promote authentic, problem solving–focused classroom environments. The shift has had a detrimental impact on student acquisition of vitally important skills, which many refer to as 21st century skills, and often the ability of students to "think." Further, schooling has become increasingly siloed into compartments of mathematics, science, English language arts, and social studies, lacking any of the connections that are overwhelmingly present in

the real world around children. Students have experienced school as content provided in boxes that must be memorized, devoid of any real-world context, and often have little understanding of why they are learning these things.

STEM-focused projects, curriculum, activities, and schools have emerged as a means to address these challenges. However, most of these efforts have continued to focus on the individual STEM disciplines (predominantly science and engineering) through more STEM classes and after-school programs in a "STEM enhanced" approach (Breiner et al. 2012). But in traditional and STEM enhanced approaches, there is little to no focus on other disciplines that are integral to the context of STEM in the real world. Integrated STEM education, on the other hand, infuses the learning of important STEM content and concepts with a much-needed emphasis on 21st century skills and a problem- and project-based pedagogy that more closely mirrors the real-world setting for society's challenges. It incorporates social studies, English language arts, and the arts as pivotal and necessary (Johnson 2013; Rennie, Venville, and Wallace 2012; Roehrig et al. 2012).

FRAMEWORK FOR STEM INTEGRATION IN THE CLASSROOM

The *STEM Road Map Curriculum Series* is grounded in the Framework for STEM Integration in the Classroom as conceptualized by Moore, Guzey, and Brown (2014) and Moore et al. (2014). The framework has six elements, described in the context of how they are used in the *STEM Road Map Curriculum Series* as follows:

1. The STEM Road Map contexts are meaningful to students and provide motivation to engage with the content. Together, these allow students to have different ways to enter into the challenge.

2. The STEM Road Map modules include engineering design that allows students to design technologies (i.e., products that are part of the designed world) for a compelling purpose.

3. The STEM Road Map modules provide students with the opportunities to learn from failure and redesign based on the lessons learned.

4. The STEM Road Map modules include standards-based disciplinary content as the learning objectives.

5. The STEM Road Map modules include student-centered pedagogies that allow students to grapple with the content, tie their ideas to the context, and learn to think for themselves as they deepen their conceptual knowledge.

6. The STEM Road Map modules emphasize 21st century skills and, in particular, highlight communication and teamwork.

All of the STEM Road Map modules incorporate these six elements; however, the level of emphasis on each of these elements varies based on the challenge or problem in each module.

THE NEED FOR THE *STEM ROAD MAP CURRICULUM SERIES*

As focus is increasing on integrated STEM, and additional schools and programs decide to move their curriculum and instruction in this direction, there is a need for high-quality, research-based curriculum designed with integrated STEM at the core. Several good resources are available to help teachers infuse engineering or more STEM enhanced approaches, but no curriculum exists that spans K–12 with an integrated STEM focus. The next chapter provides detailed information about the specific pedagogy, instructional strategies, and learning theory on which the *STEM Road Map Curriculum Series* is grounded.

REFERENCES

Breiner, J., M. Harkness, C. C. Johnson, and C. Koehler. 2012. What is STEM? A discussion about conceptions of STEM in education and partnerships. *School Science and Mathematics* 112 (1): 3–11.

Johnson, C. C. 2013. Conceptualizing integrated STEM education: Editorial. *School Science and Mathematics* 113 (8): 367–368.

Koehler, C. M., M. A. Bloom, and I. C. Binns. 2013. Lights, camera, action: Developing a methodology to document mainstream films' portrayal of nature of science and scientific inquiry. *Electronic Journal of Science Education* 17 (2).

Moore, T. J., S. S. Guzey, and A. Brown. 2014. Greenhouse design to increase habitable land: An engineering unit. *Science Scope* 51–57.

Moore, T. J., M. S. Stohlmann, H.-H. Wang, K. M. Tank, A. W. Glancy, and G. H. Roehrig. 2014. Implementation and integration of engineering in K–12 STEM education. In *Engineering in pre-college settings: Synthesizing research, policy, and practices,* ed. S. Purzer, J. Strobel, and M. Cardella, 35–60. West Lafayette, IN: Purdue Press.

Rennie, L., G. Venville, and J. Wallace. 2012. *Integrating science, technology, engineering, and mathematics: Issues, reflections, and ways forward.* New York: Routledge.

Roehrig, G. H., T. J. Moore, H. H. Wang, and M. S. Park. 2012. Is adding the *E* enough? Investigating the impact of K–12 engineering standards on the implementation of STEM integration. *School Science and Mathematics* 112 (1): 31–44.

2

STRATEGIES USED IN THE *STEM ROAD MAP CURRICULUM SERIES*

Erin Peters-Burton, Carla C. Johnson, Toni A. Sondergeld, and Tamara J. Moore

The *STEM Road Map Curriculum Series* uses what has been identified through research as best-practice pedagogy, including embedded formative assessment strategies throughout each module. This chapter briefly describes the key strategies that are employed in the series.

PROJECT- AND PROBLEM-BASED LEARNING

Each module in the *STEM Road Map Curriculum Series* uses either project-based learning or problem-based learning to drive the instruction. Project-based learning begins with a driving question to guide student teams in addressing a contextualized local or community problem or issue. The outcome of project-based instruction is a product that is conceptualized, designed, and tested through a series of scaffolded learning experiences (Blumenfeld et al. 1991; Krajcik and Blumenfeld 2006). Problem-based learning is often grounded in a fictitious scenario, challenge, or problem (Barell 2006; Lambros 2004). On the first day of instruction within the unit, student teams are provided with the context of the problem. Teams work through a series of activities and use open-ended research to develop their potential solution to the problem or challenge, which need not be a tangible product (Johnson 2003).

ENGINEERING DESIGN PROCESS

The *STEM Road Map Curriculum Series* uses engineering design as a way to facilitate integrated STEM within the modules. The engineering design process (EDP) is depicted in Figure 2.1 (p. 10). It highlights two major aspects of engineering design—problem scoping and solution generation—and six specific components of working toward a design: define the problem, learn about the problem, plan a solution, try the solution, test the solution, decide whether the solution is good enough. It also shows that communication

Figure 2.1. Engineering Design Process

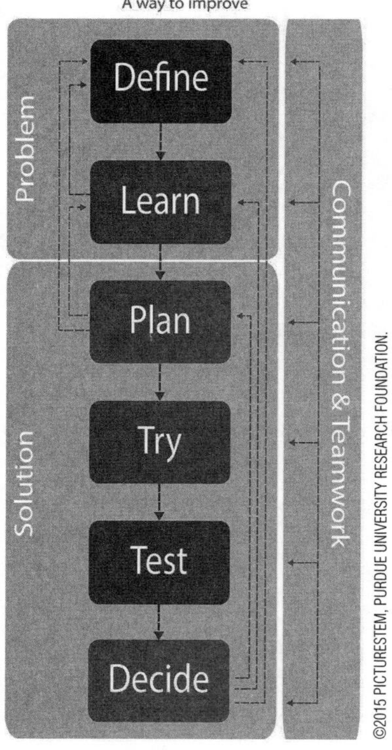

and teamwork are involved throughout the entire process. As the arrows in the figure indicate, the order in which the components of engineering design are addressed depends on what becomes needed as designers progress through the EDP. Designers must communicate and work in teams throughout the process. The EDP is iterative, meaning that components of the process can be repeated as needed until the design is good enough to present to the client as a potential solution to the problem.

Problem scoping is the process of gathering and analyzing information to deeply understand the engineering design problem. It includes defining the problem and learning about the problem. Defining the problem includes identifying the problem, the client, and the end user of the design. The client is the person (or people) who hired the designers to do the work, and the end user is the person (or people) who will use the final design. The designers must also identify the criteria and the constraints of the problem. The criteria are the things the client wants from the solution, and the constraints are the things that limit the possible solutions. The designers must spend significant time learning about the problem, which can include activities such as the following:

- Reading informational texts and researching about relevant concepts or contexts
- Identifying and learning about needed mathematical and scientific skills, knowledge, and tools
- Learning about things done previously to solve similar problems
- Experimenting with possible materials that could be used in the design

Problem scoping also allows designers to consider how to measure the success of the design in addressing specific criteria and staying within the constraints over multiple iterations of solution generation.

Solution generation includes planning a solution, trying the solution, testing the solution, and deciding whether the solution is good enough. Planning the solution includes generating many design ideas that both address the criteria and meet the constraints. Here the designers must consider what was learned about the problem during problem scoping. Design plans include clear communication of design ideas through media such as notebooks, blueprints, schematics, or storyboards. They also include details about the design, such as measurements, materials, colors, costs of materials, instructions for how things fit together, and sets of directions. Making the decision about which design idea to move forward involves considering the trade-offs of each design idea.

Once a clear design plan is in place, the designers must try the solution. Trying the solution includes developing a prototype (a testable model) based on the plan generated. The prototype might be something physical or a process to accomplish a goal. This component of design requires that the designers consider the risk involved in implementing the design. The prototype developed must be tested. Testing the solution includes conducting fair tests that verify whether the plan is a solution that is good enough to meet the client and end user needs and wants. Data need to be collected about the results of the tests of the prototype, and these data should be used to make evidence-based decisions regarding the design choices made in the plan. Here, the designers must again consider the criteria and constraints for the problem.

Using the data gathered from the testing, the designers must decide whether the solution is good enough to meet the client and end user needs and wants by assessment based on the criteria and constraints. Here, the designers must justify or reject design decisions based on the background research gathered while learning about the problem and on the evidence gathered during the testing of the solution. The designers must now decide whether to present the current solution to the client as a possibility or to do more iterations of design on the solution. If they decide that improvements need to be made to the solution, the designers must decide if there is more that needs to be understood about the problem, client, or end user; if another design idea should be tried; or if more planning needs to be conducted on the same design. One way or another, more work needs to be done.

Throughout the process of designing a solution to meet a client's needs and wants, designers work in teams and must communicate to each other, the client, and likely the end user. Teamwork is important in engineering design because multiple perspectives and differing skills and knowledge are valuable when working to solve problems. Communication is key to the success of the designed solution. Designers must communicate their ideas clearly using many different representations, such as text in an engineering notebook, diagrams, flowcharts, technical briefs, or memos to the client.

LEARNING CYCLE

The same format for the learning cycle is used in all grade levels throughout the STEM Road Map, so that students engage in a variety of activities to learn about phenomena in the modules thoroughly and have consistent experiences in the problem- and project-based learning modules. Expectations for learning by younger students are not as high as for older students, but the format of the progression of learning is the same. Students who have learned with curriculum from the STEM Road Map in early grades know

what to expect in later grades. The learning cycle consists of five parts—Introductory Activity/Engagement, Activity/Exploration, Explanation, Elaboration/Application of Knowledge, and Evaluation/Assessment—and is based on the empirically tested 5E model from BSCS (Bybee et al. 2006).

In the Introductory Activity/Engagement phase, teachers introduce the module challenge and use a unique approach designed to pique students' curiosity. This phase gets students to start thinking about what they already know about the topic and begin wondering about key ideas. The Introductory Activity/Engagement phase positions students to be confident about what they are about to learn, because they have prior knowledge, and clues them into what they don't yet know.

In the Activity/Exploration phase, the teacher sets up activities in which students experience a deeper look at the topics that were introduced earlier. Students engage in the activities and generate new questions or consider possibilities using preliminary investigations. Students work independently, in small groups, and in whole-group settings to conduct investigations, resulting in common experiences about the topic and skills involved in the real-world activities. Teachers can assess students' development of concepts and skills based on the common experiences during this phase.

During the Explanation phase, teachers direct students' attention to concepts they need to understand and skills they need to possess to accomplish the challenge. Students participate in activities to demonstrate their knowledge and skills to this point, and teachers can pinpoint gaps in student knowledge during this phase.

In the Elaboration/Application of Knowledge phase, teachers present students with activities that engage in higher-order thinking to create depth and breadth of student knowledge, while connecting ideas across topics within and across STEM. Students apply what they have learned thus far in the module to a new context or elaborate on what they have learned about the topic to a deeper level of detail.

In the last phase, Evaluation/Assessment, teachers give students summative feedback on their knowledge and skills as demonstrated through the challenge. This is not the only point of assessment (as discussed in the section on Embedded Formative Assessments), but it is an assessment of the culmination of the knowledge and skills for the module. Students demonstrate their cognitive growth at this point and reflect on how far they have come since the beginning of the module. The challenges are designed to be multidimensional in the ways students must collaborate and communicate their new knowledge.

STEM RESEARCH NOTEBOOK

One of the main components of the *STEM Road Map Curriculum Series* is the STEM Research Notebook, a place for students to capture their ideas, questions, observations, reflections, evidence of progress, and other items associated with their daily work. At the beginning of each module, the teacher walks students through the setup of the STEM Research Notebook, which could be a three-ring binder, composition book, or spiral notebook. You may wish to have students create divided sections so that they can easily access work from various disciplines during the module. Electronic notebooks kept on student devices are also acceptable and encouraged. Students will develop their own table of contents and create chapters in the notebook for each module.

Each lesson in the *STEM Road Map Curriculum Series* includes one or more prompts that are designed for inclusion in the STEM Research Notebook and appear as questions or statements that the teacher assigns to students. These prompts require students to apply what they have learned across the lesson to solve the big problem or challenge for that module. Each lesson is designed to meaningfully refer students to the larger problem or challenge they have been assigned to solve with their teams. The STEM Research Notebook is designed to be a key formative assessment tool, as students' daily entries provide evidence of what they are learning. The notebook can be used as a mechanism for dialogue between the teacher and students, as well as for peer and self-evaluation.

The use of the STEM Research Notebook is designed to scaffold student notebooking skills across the grade bands in the *STEM Road Map Curriculum Series*. In the early grades, children learn how to organize their daily work in the notebook as a way to collect their products for future reference. In elementary school, students structure their notebooks to integrate background research along with their daily work and lesson prompts. In the upper grades (middle and high school), students expand their use of research and data gathering through team discussions to more closely mirror the work of STEM experts in the real world.

THE ROLE OF ASSESSMENT IN THE *STEM ROAD MAP CURRICULUM SERIES*

Starting in the middle years and continuing into secondary education, the word *assessment* typically brings grades to mind. These grades may take the form of a letter or a percentage, but they typically are used as a representation of a student's content mastery. If well thought out and implemented, however, classroom assessment can offer teachers, parents, and students valuable information about student learning and misconceptions that does not necessarily come in the form of a grade (Popham 2013).

The *STEM Road Map Curriculum Series* provides a set of assessments for each module. Teachers are encouraged to use assessment information for more than just assigning grades to students. Instead, assessments of activities requiring students to actively engage in their learning, such as student journaling in STEM Research Notebooks, collaborative presentations, and constructing graphic organizers, should be used to move student learning forward. Whereas other curriculum with assessments may include objective-type (multiple-choice or matching) tests, quizzes, or worksheets, we have intentionally avoided these

forms of assessments to better align assessment strategies with teacher instruction and student learning techniques. Since the focus of this book is on project- or problem-based STEM curriculum and instruction that focuses on higher-level thinking skills, appropriate and authentic performance assessments were developed to elicit the most reliable and valid indication of growth in student abilities (Brookhart and Nitko 2008).

Comprehensive Assessment System

Assessment throughout all STEM Road Map curriculum modules acts as a comprehensive system in which formative and summative assessments work together to provide teachers with high-quality information on student learning. Formative assessment occurs when the teacher finds out formally or informally what a student knows about a smaller, defined concept or skill and provides timely feedback to the student about his or her level of proficiency. Summative assessments occur when students have performed all activities in the module and are given a cumulative performance evaluation in which they demonstrate their growth in learning.

A comprehensive assessment system can be thought of as akin to a sporting event. Formative assessments are the practices: It is important to accomplish them consistently, they provide feedback to help students improve their learning, and making mistakes can be worthwhile if students are given an opportunity to learn from them. Summative assessments are the competitions: Students need to be prepared to perform at the best of their ability. Without multiple opportunities to practice skills along the way through formative assessments, students will not have the best chance of demonstrating growth in abilities through summative assessments (Black and Wiliam 1998).

Embedded Formative Assessments

Formative assessments in this module serve two main purposes: to provide feedback to students about their learning and to provide important information for the teacher to inform immediate instructional needs. Providing feedback to students is particularly important when conducting problem- or project-based learning because students take on much of the responsibility for learning, and teachers must facilitate student learning in an informed way. For example, if students are required to conduct research for the Activity/Exploration phase but are not familiar with what constitutes a reliable resource, they may develop misconceptions based on poor information. When a teacher monitors this learning through formative assessments and provides specific feedback related to the instructional goals, students are less likely to develop incomplete or incorrect conceptions in their independent investigations. By using formative assessment to detect problems in student learning and then acting on this information, teachers help move student learning forward through these teachable moments.

Formative assessments come in a variety of formats. They can be informal, such as asking students probing questions related to student knowledge or tasks or simply observing students engaged in an activity to gather information about student skills. Formative assessments can also be formal, such as a written quiz or a laboratory practical. Regardless of the type, three key steps must be completed when using formative assessments (Sondergeld, Bell, and Leusner 2010). First, the assessment is delivered to students so that teachers can collect data. Next, teachers analyze the data (student responses) to determine student strengths and areas that need additional support. Finally, teachers use the results from information collected to modify lessons and create learning environments that reinforce weak points in student learning. If student learning information is not used to modify instruction, the assessment cannot be considered formative in nature.

Formative assessments can be about content, science process skills, or even learning skills. When a formative assessment focuses on content, it assesses student knowledge about the disciplinary core ideas from the *Next Generation Science Standards* (*NGSS*) or content objectives from *Common Core State Standards for Mathematics* (*CCSS Mathematics*) or *Common Core State Standards for English Language Arts* (*CCSS ELA*). Content-focused formative assessments ask students questions about declarative knowledge regarding the concepts they have been learning. Process skills formative assessments examine the extent to which a student can perform science and engineering practices from the *NGSS* or process objectives from *CCSS Mathematics* or *CCSS ELA*, such as constructing an argument. Learning skills can also be assessed formatively by asking students to reflect on the ways they learn best during a module and identify ways they could have learned more.

Assessment Maps

Assessment maps or blueprints can be used to ensure alignment between classroom instruction and assessment. If what students are learning in the classroom is not the same as the content on which they are assessed, the resultant judgment made on student learning will be invalid (Brookhart and Nitko 2008). Therefore, the issue of instruction and assessment alignment is critical. The assessment map for this book (found in Chapter 3) indicates by lesson whether the assessment should be completed as a group or on an individual basis, identifies the assessment as formative or summative in nature, and aligns the assessment with its corresponding learning objectives.

Note that the module includes far more formative assessments than summative assessments. This is done intentionally to provide students with multiple opportunities to practice their learning of new skills before completing a summative assessment. Note also that formative assessments are used to collect information on only one or two learning objectives at a time so that potential relearning or instructional modifications can focus on smaller and more manageable chunks of information. Conversely, summative assessments in the module cover many more learning objectives, as they are traditionally used as final

markers of student learning. This is not to say that information collected from summative assessments cannot or should not be used formatively. If teachers find that gaps in student learning persist after a summative assessment is completed, it is important to revisit these existing misconceptions or areas of weakness before moving on (Black et al. 2003).

SELF-REGULATED LEARNING THEORY IN THE STEM ROAD MAP MODULES

Many learning theories are compatible with the STEM Road Map modules, such as constructivism, situated cognition, and meaningful learning. However, we feel that the self-regulated learning theory (SRL) aligns most appropriately (Zimmerman 2000). SRL requires students to understand that thinking needs to be motivated and managed (Ritchhart, Church, and Morrison 2011). The STEM Road Map modules are student centered and are designed to provide students with choices, concrete hands-on experiences, and opportunities to see and make connections, especially across subjects (Eliason and Jenkins 2012; NAEYC 2016). Additionally, SRL is compatible with the modules because it fosters a learning environment that supports students' motivation, enables students to become aware of their own learning strategies, and requires reflection on learning while experiencing the module (Peters and Kitsantas 2010).

Figure 2.2. SRL Theory

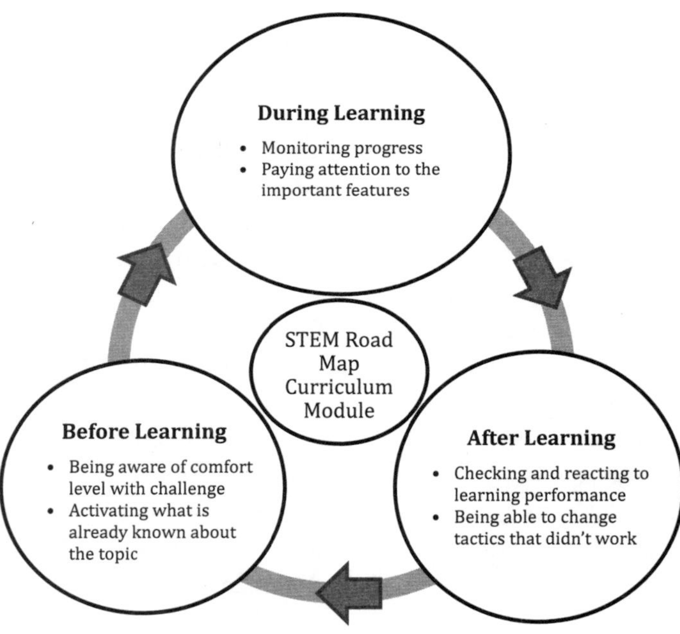

Source: Adapted from Zimmerman 2000.

The theory behind SRL (see Figure 2.2) explains the different processes that students engage in before, during, and after a learning task. Because SRL is a cyclical learning process, the accomplishment of one cycle develops strategies for the next learning cycle. This cyclic way of learning aligns with the various sections in the STEM Road Map lesson plans on Introductory Activity/Engagement, Activity/Exploration, Explanation, Elaboration/Application of Knowledge, and Evaluation/Assessment. Since the students engaged in a module take on much of the responsibility for learning, this theory also provides guidance for teachers to keep students on the right track.

The remainder of this section explains how SRL theory is embedded within the five sections of each module and points out ways to support students in becoming independent learners of STEM while productively functioning in collaborative teams.

Before Learning: Setting the Stage

Before attempting a learning task such as the STEM Road Map modules, teachers should develop an understanding of their students' level of comfort with the process of accomplishing the learning and determine what they already know about the topic. When students are comfortable with attempting a learning task, they tend to take more risks in learning and as a result achieve deeper learning (Bandura 1986).

The STEM Road Map curriculum modules are designed to foster excitement from the very beginning. Each module has an Introductory Activity/Engagement section that introduces the overall topic from a unique and exciting perspective, engaging the students to learn more so that they can accomplish the challenge. The Introductory Activity also has a design component that helps teachers assess what students already know about the topic of the module. In addition to the deliberate designs in the lesson plans to support SRL, teachers can support a high level of student comfort with the learning challenge by finding out if students have ever accomplished the same kind of task and, if so, asking them to share what worked well for them.

During Learning: Staying the Course

Some students fear inquiry learning because they aren't sure what to do to be successful (Peters 2010). However, the STEM Road Map curriculum modules are embedded with tools to help students pay attention to knowledge and skills that are important for the learning task and to check student understanding along the way. One of the most important processes for learning is the ability for learners to monitor their own progress while performing a learning task (Peters 2012). The modules allow students to monitor their progress with tools such as the STEM Research Notebooks, in which they record what they know and can check whether they have acquired a complete set of knowledge and skills. The STEM Road Map modules support inquiry strategies that include previewing, questioning, predicting, clarifying, observing, discussing, and journaling (Morrison and Milner 2014). Through the use of technology throughout the modules, inquiry is supported by providing students access to resources and data while enabling them to process information, report the findings, collaborate, and develop 21st century skills.

It is important for teachers to encourage students to have an open mind about alternative solutions and procedures (Milner and Sondergeld 2015) when working through the STEM Road Map curriculum modules. Novice learners can have difficulty knowing what to pay attention to and tend to treat each possible avenue for information as equal (Benner 1984). Teachers are the mentors in a classroom and can point out ways

for students to approach learning during the Activity/Exploration, Explanation, and Elaboration/Application of Knowledge portions of the lesson plans to ensure that students pay attention to the important concepts and skills throughout the module. For example, if a student is to demonstrate conceptual awareness of motion when working on roller coaster research, but the student has misconceptions about motion, the teacher can step in and redirect student learning.

After Learning: Knowing What Works

The classroom is a busy place, and it may often seem that there is no time for self-reflection on learning. Although skipping this reflective process may save time in the short term, it reduces the ability to take into account things that worked well and things that didn't so that teaching the module may be improved next time. In the long run, SRL skills are critical for students to become independent learners who can adapt to new situations. By investing the time it takes to teach students SRL skills, teachers can save time later, because students will be able to apply methods and approaches for learning that they have found effective to new situations. In the Evaluation/Assessment portion of the STEM Road Map curriculum modules, as well as in the formative assessments throughout the modules, two processes in the after-learning phase are supported: evaluating one's own performance and accounting for ways to adapt tactics that didn't work well. Students have many opportunities to self-assess in formative assessments, both in groups and individually, using the rubrics provided in the modules.

The designs of the *NGSS* and *CCSS* allow for students to learn in diverse ways, and the STEM Road Map curriculum modules emphasize that students can use a variety of tactics to complete the learning process. For example, students can use STEM Research Notebooks to record what they have learned during the various research activities. Notebook entries might include putting objectives in students' own words, compiling their prior learning on the topic, documenting new learning, providing proof of what they learned, and reflecting on what they felt successful doing and what they felt they still needed to work on. Perhaps students didn't realize that they were supposed to connect what they already knew with what they learned. They could record this and would be prepared in the next learning task to begin connecting prior learning with new learning.

SAFETY IN STEM

Student safety is a primary consideration in all subjects but is an area of particular concern in science, where students may interact with unfamiliar tools and materials that may pose additional safety risks. It is important to implement safety practices within the context of STEM investigations, whether in a classroom laboratory or in the field. When you keep safety in mind as a teacher, you avoid many potential issues with the lesson while also protecting your students.

STEM safety practices encompass things considered in the typical science classroom. Ensure that students are familiar with basic safety considerations, such as wearing protective equipment (e.g., safety glasses or goggles and latex-free gloves) and taking care with sharp objects, and know emergency exit procedures. Teachers should learn beforehand the locations of the safety eyewash, fume hood, fire extinguishers, and emergency shut-off switch in the classroom and how to use them. Also be aware of any school or district safety policies that are in place and apply those that align with the work being conducted in the lesson. It is important to review all safety procedures annually.

STEM investigations should always be supervised. Each lesson in the modules includes teacher guidelines for applicable safety procedures that should be followed. Before each investigation, teachers should go over these safety procedures with the student teams. Some STEM focus areas such as engineering require that students can demonstrate how to properly use equipment in the maker space before the teacher allows them to proceed with the lesson.

Information about classroom science safety, including a safety checklist for science classrooms, general lab safety recommendations, and links to other science safety resources, is available at the Council of State Science Supervisors (CSSS) website at *www.csss-science.org/safety.shtml*. The National Science Teachers Association (NSTA) provides a list of science rules and regulations, including standard operating procedures for lab safety, and a safety acknowledgement form for students and parents or guardians to sign. You can access these resources at *http://static.nsta.org/pdfs/SafetyInTheScienceClassroom.pdf*. In addition, NSTA's Safety in the Science Classroom web page (*www.nsta.org/safety*) has numerous links to safety resources, including papers written by the NSTA Safety Advisory Board.

Disclaimer: The safety precautions for each activity are based on use of the recommended materials and instructions, legal safety standards, and better professional practices. Using alternative materials or procedures for these activities may jeopardize the level of safety and therefore is at the user's own risk.

REFERENCES

Bandura, A. 1986. *Social foundations of thought and action: A social cognitive theory*. Englewood Cliffs, NJ: Prentice-Hall.

Barell, J. 2006. *Problem-based learning: An inquiry approach*. Thousand Oaks, CA: Corwin Press.

Benner, P. 1984. *From novice to expert: Excellence and power in clinical nursing practice*. Menlo Park, CA: Addison-Wesley Publishing Company.

Black, P., C. Harrison, C. Lee, B. Marshall, and D. Wiliam. 2003. *Assessment for learning: Putting it into practice*. Berkshire, UK: Open University Press.

Black, P., and D. Wiliam. 1998. Inside the black box: Raising standards through classroom assessment. *Phi Delta Kappan* 80 (2): 139–148.

Blumenfeld, P., E. Soloway, R. Marx, J. Krajcik, M. Guzdial, and A. Palincsar. 1991. Motivating project-based learning: Sustaining the doing, supporting learning. *Educational Psychologist* 26 (3): 369–398.

Brookhart, S. M., and A. J. Nitko. 2008. *Assessment and grading in classrooms*. Upper Saddle River, NJ: Pearson.

Bybee, R., J. Taylor, A. Gardner, P. Van Scotter, J. Carlson, A. Westbrook, and N. Landes. 2006. *The BSCS 5E instructional model: Origins and effectiveness. http://science.education.nih.gov/houseofreps.nsf/b82d55fa138783c2852572c9004f5566/$FILE/Appendix?D.pdf.*

Eliason, C. F., and L. T. Jenkins. 2012. *A practical guide to early childhood curriculum*. 9th ed. New York: Merrill.

Johnson, C. 2003. Bioterrorism is real-world science: Inquiry-based simulation mirrors real life. *Science Scope* 27 (3): 19–23.

Krajcik, J., and P. Blumenfeld. 2006. Project-based learning. In *The Cambridge handbook of the learning sciences*, ed. R. Keith Sawyer, 317–334. New York: Cambridge University Press.

Lambros, A. 2004. *Problem-based learning in middle and high school classrooms: A teacher's guide to implementation*. Thousand Oaks, CA: Corwin Press.

Milner, A. R., and T. Sondergeld. 2015. Gifted urban middle school students: The inquiry continuum and the nature of science. *National Journal of Urban Education and Practice* 8 (3): 442–461.

Morrison, V., and A. R. Milner. 2014. Literacy in support of science: A closer look at cross-curricular instructional practice. *Michigan Reading Journal* 46 (2): 42–56.

National Association for the Education of Young Children (NAEYC). 2016. Developmentally appropriate practice position statements. *www.naeyc.org/positionstatements/dap.*

Peters, E. E. 2010. Shifting to a student-centered science classroom: An exploration of teacher and student changes in perceptions and practices. *Journal of Science Teacher Education* 21 (3): 329–349.

Peters, E. E. 2012. Developing content knowledge in students through explicit teaching of the nature of science: Influences of goal setting and self-monitoring. *Science and Education* 21 (6): 881–898.

Peters, E. E., and A. Kitsantas. 2010. The effect of nature of science metacognitive prompts on science students' content and nature of science knowledge, metacognition, and self-regulatory efficacy. *School Science and Mathematics* 110: 382–396.

Popham, W. J. 2013. *Classroom assessment: What teachers need to know*. 7th ed. Upper Saddle River, NJ: Pearson.

Ritchhart, R., M. Church, and K. Morrison. 2011. *Making thinking visible: How to promote engagement, understanding, and independence for all learners*. San Francisco, CA: Jossey-Bass.

Sondergeld, T. A., C. A. Bell, and D. M. Leusner. 2010. Understanding how teachers engage in formative assessment. *Teaching and Learning* 24 (2): 72–86.

Zimmerman, B. J. 2000. Attaining self-regulation: A social-cognitive perspective. In *Handbook of self-regulation*, ed. M. Boekaerts, P. Pintrich, and M. Zeidner, 13–39. San Diego: Academic Press.

PACKAGING DESIGN

STEM ROAD MAP MODULE

PACKAGING DESIGN MODULE OVERVIEW

Adrienne Redmond-Sanogo, Sue Christian Parsons, Janet B. Walton, Carla C. Johnson, Erin Peters-Burton, Juliana Utley, and Toni A. Ivey

THEME: The Represented World

LEAD DISCIPLINES: Mathematics and English Language Arts

MODULE SUMMARY

Over the past decade, human ability to communicate through the use of technology has grown exponentially. Adolescents are engaged in communicating every day via texting or social media such as Twitter, Facebook, and Instagram, sometimes without one spoken word. In the Packaging Design module, students explore the realm of communication. English language arts and mathematics teachers take the lead in this unit, integrating with science and social studies contexts, which could be collaborations with these classes. In this 25-day extended three-lesson module, students explore packaging—in particular, nested packaging—to repurpose a product or market a product to a new user. As they explore, they develop both content knowledge and strong written and verbal communication skills. Persuasive writing is emphasized in this module, as students have to try to convince a client that their new product is marketable. As the students think about nested packaging, they develop understanding of geometric properties of three -dimensional shapes and engineering design. Learners' success in the 21st century workplace and beyond hinges on their ability to meld communication skills with content skills (adapted from Johnson et al. 2015, p. 100).

ESTABLISHED GOALS AND OBJECTIVES

At the conclusion of this module, students will be able to do the following:

- Explain how companies purposely target specific audiences to maximize profits when creating, designing, and marketing products
- Understand problem-solution text structures, and use that strategy to understand authentic literature

- Understand the role of demographics in packaging and marketing, and develop a demographic profile for their products
- Understand where products and packaging originate and end up (life cycle of a product)
- Discuss the sustainability issues associated with packaging and manufacturing of products
- Understand how surface area and volume are used in packaging and manufacturing
- Calculate surface area and volume of three-dimensional figures, and develop a general formula
- Understand that marketing is a complex process that requires feedback from a target market and revisions as needed
- Understand that statistics can be misleading and that it is the job of the consumer to fact check statistics
- Understand that all media messages are constructed, and when engaging with a media message, consider who created it and for what purpose
- Understand that media messages are constructed using a creative language with its own rules, and when engaging with a media message, consider what techniques are being employed to attract buyer attention
- Understand that different people experience the same media message differently, and consider their impressions and how others might view the message differently
- Understand that media have embedded values and points of view, and consider what values and points of views are represented
- Understand that most media messages are designed to gain profit or power, and consider why a message was sent and how an effective marketing message is created
- Select and use multiple forms of media (visual and textual) to convey information about a product and persuade an audience to buy it
- Understand the role that the economy plays in society

CHALLENGE OR PROBLEM FOR STUDENTS TO SOLVE: PRODUCT DESIGN CHALLENGE

In this design project, students are challenged to reimagine and develop a new way of packaging a current product on the market. This project can be launched in any of the content classes: English language arts, mathematics, science, or social studies.

CONTENT STANDARDS ADDRESSED IN THIS STEM ROAD MAP MODULE

A full listing with descriptions of the standards this module addresses can be found in the appendix. Listings of the particular standards addressed within lessons are provided in a table for each lesson in Chapter 4.

STEM RESEARCH NOTEBOOK

Each student should maintain a STEM Research Notebook, which will serve as a place for students to organize their work throughout this module (see p. 12 for more general discussion on setup and use of this notebook). All written work in the module should be included in the notebook, including records of students' thoughts and ideas, fictional accounts based on the concepts in the module, and records of student progress through the engineering design process (EDP). The notebooks may be maintained across subject areas, giving students the opportunity to see that although their classes may be separated during the school day, the knowledge they gain is connected.

Each lesson in this module includes student handouts that should be kept in the STEM Research Notebooks after completion, as well as a prompt to which students should respond in their notebooks. You may also wish to have students include the STEM Research Notebook Guidelines student handout on page 26 in their notebooks.

Emphasize to students the importance of organizing all information in a Research Notebook. Explain to them that scientists and other researchers maintain detailed Research Notebooks in their work. These notebooks, which are crucial to researchers' work because they contain critical information and track the researchers' progress, are often considered legal documents for scientists who are pursuing patents or wish to provide proof of their discovery process.

STUDENT HANDOUT

STEM RESEARCH NOTEBOOK GUIDELINES

STEM professionals record their ideas, inventions, experiments, questions, observations, and other work details in notebooks so that they can use these notebooks to help them think about their projects and the problems they are trying to solve. You will each keep a STEM Research Notebook during this module that is like the notebooks that STEM professionals use. In this notebook, you will include all your work and notes about ideas you have. The notebook will help you connect your daily work with the big problem or challenge you are working to solve.

It is important that you organize your notebook entries under the following headings:

1. **Chapter Topic or Title of Problem or Challenge:** You will start a new chapter in your STEM Research Notebook for each new module. This heading is the topic or title of the big problem or challenge that your team is working to solve in this module.
2. **Date and Topic of Lesson Activity for the Day:** Each day, you will begin your daily entry by writing the date and the day's lesson topic at the top of a new page. Write the page number both on the page and in the table of contents.
3. **Information Gathered From Research:** This is information you find from outside resources such as websites or books.
4. **Information Gained From Class or Discussions With Team Members:** This information includes any notes you take in class and notes about things your team discusses. You can include drawings of your ideas here, too.
5. **New Data Collected From Investigations:** This includes data gathered from experiments, investigations, and activities in class.
6. **Documents:** These are handouts and other resources you may receive in class that will help you solve your big problem or challenge. Paste or staple these documents in your STEM Research Notebook for safekeeping and easy access later.
7. **Personal Reflections:** Here, you record your own thoughts and ideas on what you are learning.
8. **Lesson Prompts:** These are questions or statements that your teacher assigns you within each lesson to help you solve your big problem or challenge. You will respond to the prompts in your notebook.
9. **Other Items:** This section includes any other items your teacher gives you or other ideas or questions you may have.

MODULE LAUNCH

This module can be launched in any of the content classes. Ideally, the schedule could be adjusted so that the team of content teachers could introduce the project together. You could also launch the module by producing a video involving all the teachers and then showing and discussing this video in the various classes.

Before students enter the room, create a display of products with nested construction and visually engaging packaging, including examples that demonstrate the kinds of product packaging they will seek to create. As much as possible, showcase the items for dramatic effect, using such things as labels, lighting, and background music. As students enter the room, hand each a name badge and welcome him or her to the Product Design Challenge. Once all your students have gathered, invite them to visit the display with the following question in mind: What aspects of the products displayed might make them appealing to a buyer? Then, divide the students into teams and provide each team with a copy of the Product Design Challenge Guidelines (pp. 93–94).

PREREQUISITE SKILLS FOR THE MODULE

Students enter this module with a wide range of preexisting skills, information, and knowledge. Table 3.1 provides an overview of prerequisite skills and knowledge that students are expected to apply in this module, along with examples of how they apply this knowledge throughout the module. Differentiation strategies are also provided for students who may need additional support in acquiring or applying this knowledge.

Table 3.1. Prerequisite Key Knowledge and Examples of Applications and Differentiation Strategies

Prerequisite Key Knowledge	Application of Knowledge	Differentiation for Students Needing Knowledge
Science: • Identify materials based on their properties through observations and measurements. • Understand the movement of matter among plants, animals, decomposers, and the environment. • Understand the ways individual communities use science ideas to protect Earth's resources and environment. • Can represent data in graphic displays to reveal patterns in data. • Have explored engineering design and can define simple problems, generate and compare multiple solutions, and plan and carry out fair tests.	Science: • Use observation and measurement skills. • Explore sustainability. • Represent and graph data to answer questions. • Use the engineering design process to solve complex problems.	Science: • Students who struggle with measurement or observation skills may need to work with partners. • Provide some technology tools such as digital thermometers to help students who are unable to use instruments to measure. • Provide students with technology resources to produce graphic representations of data. • Scaffold instruction to support students who have had little experience with engineering design.
Mathematics: • Solve challenging problems. • Use fractions and decimals. • Display data in both whole number and fractional units to solve problems.	Mathematics: • Solve real-world problems. • Learn to use percentages in real-world problems.	Mathematics: • Scaffold lessons to support student problem solving. • Students who struggle with fraction concepts can use calculators and physical models to solve problems.

Continued

Table 3.1. (*continued*)

Prerequisite Key Knowledge	Application of Knowledge	Differentiation for Students Needing Knowledge
Mathematics (*continued*): • Understand concepts of area of various two-dimensional shapes. • Understand that volume is an attribute of solid figures and that unit cubes can be used to measure volume. • Understand and can name characteristics of three-dimensional shapes.	Mathematics (*continued*): • Design their own study, collect data, and use measures of center to describe the data. • Use understanding of area, perimeter, and volume to find surface area and volume of three-dimensional figures. • Explore the properties of three-dimensional shapes.	Mathematics (*continued*): • Students can use technology to represent data and explore measures of center. • Students who have not developed a conceptual understanding of area, perimeter, and volume may need support. For example, provide color tiles to students and have them build a square with a certain number of tiles. • Sorting activities are essential to help students move from level 0 of the van Hiele model of geometric thinking.
Reading: • Know the difference between fiction and nonfiction texts.	Reading: • Explore a variety of nonfiction text structures.	Reading: • None
Writing: • Able to write for a variety of purposes.	Writing: • Write blog responses and letters to a company, develop multimedia presentations, and create marketing campaigns.	Writing: • Students who struggle with writing may use speech-to-text apps to "write" their thoughts digitally.
Social Studies: • Understand proper nutrition and the difference between processed foods and fresh foods. • Understand the difference between fact and opinion in an argument. • Have had experience with filming, editing, and developing multimedia presentations. • Able to research using the internet.	Social Studies: • Exploring food deserts and food swamps. • Distinguish between fact and opinion. • Use video and presentation software to create a multimedia presentation. • Explore social justice issues and evaluate the validity of sources.	Social Studies: • If students have had limited experience with fresh produce and meats, provide examples and experiences to help them understand the difference between fresh and processed foods. • Provide students with examples of facts and opinions used in the media. • Enlist the school's technology expert to help students who are struggling with their multimedia projects or pair inexperienced students with knowledgeable peers.

POTENTIAL STEM MISCONCEPTIONS

Students enter the classroom with a wide variety of prior knowledge and ideas, so it is important to be alert to misconceptions, or inappropriate understandings of foundational knowledge. These misconceptions can be classified as one of several types: "preconceived notions," opinions based on popular beliefs or understandings; "nonscientific beliefs," knowledge students have gained about science from sources outside the scientific community; "conceptual misunderstandings," incorrect conceptual models based on incomplete understanding of concepts; "vernacular misconceptions," misunderstandings of words based on their common use versus their scientific use; and "factual misconceptions," incorrect or imprecise knowledge learned in early life that remains unchallenged (NRC 1997, p. 28). Misconceptions must be addressed and dismantled in order for students to reconstruct their knowledge, and therefore teachers should be prepared to take the following steps:

- *Identify students' misconceptions.*
- *Provide a forum for students to confront their misconceptions.*
- *Help students reconstruct and internalize their knowledge, based on scientific models.* (NRC 1997, p. 29)

Keeley and Harrington (2010) recommend using diagnostic tools such as probes and formative assessment to identify and confront student misconceptions and begin the process of reconstructing student knowledge. Keeley and Harrington's *Uncovering Student Ideas in Science* series contains probes targeted toward uncovering student misconceptions in a variety of areas and may be useful resources for addressing student misconceptions in this module.

Some commonly held misconceptions specific to lesson content are provided with each lesson so that you can be alert for student misunderstanding of the science concepts presented and used during this module. The American Association for the Advancement of Science has also identified misconceptions that students frequently hold regarding various science concepts (see the links at *http://assessment.aaas.org/topics*).

SRL PROCESS COMPONENTS

Table 3.2 illustrates some of the activities in the Package Design module and how they align to the self-regulated learning (SRL) processes before, during, and after learning.

Table 3.2. SRL Process Components

Learning Process Components	Example From Packaging Design Module	Lesson Number and Learning Component
BEFORE LEARNING		
Motivates students	Students complete the Where's My Stuff? exploration, in which they take digital pictures of their stuff and collect data.	Lesson 1, Introductory Activity/ Engagement
Evokes prior learning	Students use a familiar product, tortilla chips, to explore components of marketing and packaging.	Lesson 1, Activity/ Engagement
DURING LEARNING		
Focuses on important features	Students explore marketing and packaging from the perspective of each specific discipline: sustainability for science, shapes for mathematics, media messages for ELA, and geography for social studies. Teachers help students see the same concept through different lenses.	Lesson 2, Introductory Activity/ Engagement
Helps students monitor their progress	Students present their ideas about sustainability and life cycle of packaging to their peers. Teacher and peers help assess whether they have considered all facets of the challenge.	Lesson 2, Elaboration/ Application of Knowledge
AFTER LEARNING		
Evaluates learning	Students get feedback on their final challenge product from peers and classroom guests.	Lesson 3, Elaboration/ Application of Knowledge
Takes account of what worked and what did not work	Students reflect on the feedback they receive when they present to a panel of fictional "company executives."	Lesson 3, Elaboration/ Application of Knowledge

STRATEGIES FOR DIFFERENTIATING INSTRUCTION WITHIN THIS MODULE

For the purposes of this curriculum module, differentiated instruction is conceptualized as a way to tailor instruction—including process, content, and product—to various student needs in your class. A number of differentiation strategies are integrated into lessons across the module. The problem- and project-based learning approach used in the lessons are designed to address students' multiple intelligences by providing a variety of entry points and methods to investigate the key concepts in the module (for example, investigating packaging from the perspectives of science and social issues via scientific inquiry, literature, journaling, and collaborative design). Differentiation strategies for students needing support in prerequisite knowledge can be found in Table 3.1 (pp. 28–29). You are encouraged to use information gained about student prior knowledge during introductory activities and discussions to inform your instructional differentiation. Strategies incorporated into this lesson include flexible grouping, varied environmental learning contexts, assessments, compacting, and tiered assignments and scaffolding.

Flexible Grouping: Students work collaboratively in a variety of activities throughout this module. Grouping strategies you may choose to employ include student-led grouping, grouping students according to ability level or common interests, grouping students randomly, or grouping them so that students in each group have complementary strengths (for instance, one student might be strong in mathematics, another in art, and another in writing).

Varied Environmental Learning Contexts: Students have the opportunity to learn in various contexts throughout the module, including alone, in groups, in quiet reading and research-oriented activities, and in active learning through inquiry and design activities. In addition, students learn in a variety of ways, including through doing inquiry activities, journaling, reading a variety of texts, watching videos, participating in class discussion, and conducting web-based research.

Assessments: Students are assessed in a variety of ways throughout the module, including individual and collaborative formative and summative assessments. Students have the opportunity to produce work via written text, oral and media presentations, and modeling. You may choose to provide students with additional choices of media for their products (for example, PowerPoint presentations, posters, or student-created websites or blogs).

Compacting: Based on student prior knowledge, you may wish to adjust instructional activities for students who exhibit prior mastery of a learning objective. For instance, if some students exhibit mastery of determining arithmetic means and collecting data in Lesson 1, you may wish to limit the amount of time they spend practicing these skills and instead introduce ELA or social studies connections with associated activities.

Tiered Assignments and Scaffolding: Based on your awareness of student ability, understanding of concepts, and mastery of skills, you may wish to provide students with variations on activities by adding complexity to assignments or providing more or fewer learning supports for activities throughout the module. For instance, some students may need additional support in identifying key search words and phrases for web-based research or may benefit from cloze sentence handouts to enhance vocabulary understanding. Other students may benefit from expanded reading selections and additional reflective writing or from working with manipulatives and other visual representations of mathematical concepts. You may also work with your school librarian to compile a set of topical resources at a variety of reading levels.

STRATEGIES FOR ENGLISH LANGUAGE LEARNERS

Students who are developing proficiency in English language skills require additional supports to simultaneously learn academic content and the specialized language associated with specific content areas. WIDA (2012) has created a framework for providing support to these students and makes available rubrics and guidance on differentiating instructional materials for English language learners (ELLs) (see *www.wida.us/get.aspx?id=7*). In particular, ELL students may benefit from additional sensory supports such as images, physical modeling, and graphic representations of module content, as well as interactive support through collaborative work. Teachers differentiating instruction for ELL students should carefully consider the needs of these students as they introduce and use academic language in various language domains (listening, speaking, reading, and writing) throughout this module. To adequately differentiate instruction for ELL students, you should have an understanding of the proficiency level of each student. The following five overarching WIDA learning standards are relevant to this module:

- Standard 1: Social and Instructional language. Focus on social behavior in group work and class discussions.
- Standard 2: The language of Language Arts. Focus on forms of print, elements of text, picture books, comprehension strategies, main ideas and details, persuasive language, creation of informational text, and editing and revision.
- Standard 3: The language of Mathematics. Focus on numbers and operations, patterns, number sense, measurement, and strategies for problem solving.
- Standard 4: The language of Science. Focus on safety practices, scientific process, and scientific inquiry.
- Standard 5: The language of Social Studies. Focus on consumers and producers, resources, and environmental issues.

SAFETY CONSIDERATIONS FOR THE ACTIVITIES IN THIS MODULE

In this module, a few safety concerns must be addressed with students before beginning Lesson 1. Caution students that they should not throw chips, weights, or other objects. When using weights, they must take care not to injure their classmates or school property. If a student has a food allergy, the Save the Chips Competition may need to be modified to use fictitious bags of chips that you create. For more general safety guidelines, see the Safety in STEM section in Chapter 2 (p. 18). Internet safety is also important. In this module, students are expected to conduct internet research on a variety of topics. Develop an internet safety protocol and discuss it with students. Share the safety protocol with parents so they can monitor students' use at home as they work on aspects of their projects and conduct internet research.

DESIRED OUTCOMES AND MONITORING SUCCESS

The desired outcomes for this module are outlined in Table 3.3, along with suggested ways to gather evidence to monitor student success. For more specific details on desired outcomes, see the Established Goals and Objectives sections for the module and individual lessons.

Table 3.3. Desired Outcomes and Evidence of Success

Desired Outcome	Evidence of Success	
	PERFORMANCE TASKS	OTHER MEASURES
Students work in a team to design packaging, create a logo, and develop a marketing strategy that they present to a panel of "company executives."	Students are assessed using project rubrics that focus on content and application of skills related to academic content. Students maintain STEM Research Notebooks that contain designs, research notes, evidence of collaboration, and unit-related work from all classes.	The project rubrics have participation built in, so there are no separate measures.

ASSESSMENT PLAN OVERVIEW AND MAP

Table 3.4 provides an overview of the major group and individual products and deliverables, or things that student teams will produce in this module, that constitute the assessment for this module. See Table 3.5 for a full assessment map of formative and summative assessments in this module.

Table 3.4. Major Products and Deliverables for Groups and Individuals

Lesson	Major Group Products and Deliverables	Major Individual Products and Deliverables
1	• Save the Chips Competition multimedia presentation	• Letter to company • Response to blog post • STEM Research Notebook prompts
2	• Package type research • Life Cycle poster • Social justice presentation	• STEM Research Notebook prompts
3	• Product Design Challenge • Social media campaign	• STEM Research Notebook prompts

Table 3.5. Assessment Map for Packaging Design Module

Lesson	Assessment	Group/ Individual	Formative/ Summative	Lesson Objective Assessed
1	Save the Chip Competition	Group	Formative	• Understand that companies purposely target specific audiences to maximize profits when creating, designing, and marketing products
1	Multimedia presentation	Group	Formative	• Understand that companies purposely target specific audiences to maximize profits when creating, designing, and marketing products.
1	Letter to the company	Individual	Formative	• Understand problem-solution text structures and use that strategy to understand authentic literature.
1	Response to blog post	Individual	Formative	• Understand problem-solution text structures and use that strategy to understand authentic literature.
1	STEM Research Notebook prompts	Individual	Formative	• Explain how companies purposely target specific audiences to maximize profits when creating, designing, and marketing products. • Understand and calculate the mean as a measure of central tendency.

Continued

Table 3.5. (*continued*)

Lesson	Assessment	Group/ Individual	Formative/ Summative	Lesson Objective Assessed
2	Package type research	Group	Formative	• Understand how surface area and volume are used in packaging and manufacturing. • Calculate surface area and volume of three-dimensional figures, and develop a general formula.
2	Life Cycle poster	Group	Formative	• Understand where products and packaging originate and end up (life cycle of a product). • Understand the sustainability issues associated with packaging and manufacturing of products.
2	PowerPoint slide	Group	Formative	• Understand the sustainability issues associated with packaging and manufacturing of products.
2	Social justice presentation	Group	Formative	• Select and use a variety of media (e.g., print, art, video) to communicate complex information. • Use oral and written language effectively to collaborate and problem solve in a work community context. • Understand that companies purposely target specific audiences to maximize profits when creating, designing, and marketing products. • Understand the role of demographics in packaging and marketing, and develop a demographic profile for their product.
2	STEM Research Notebook prompts	Individual	Formative	• Understand that companies purposely target specific audiences to maximize profits when creating, designing, and marketing products. • Understand the sustainability issues associated with packaging and manufacturing of products.
3	Product Design Challenge	Group	Summative	• Understand that companies purposely target specific audiences to maximize profits when creating, designing, and marketing products. • Understand that marketing is a complex process that requires feedback from a target market and revisions as needed. • Select and use multiple forms of media (visual and textual) to convey information about a product and persuade an audience to buy it.

Continued

Table 3.5. *(continued)*

Lesson	Assessment	Group/ Individual	Formative/ Summative	Lesson Objective Assessed
3	Social media campaign	Group	Summative	• Understand that all media messages are constructed and that when engaging with a media message, it is important to consider who created it and for what purpose. • Understand that media messages are constructed using a creative language with its own rules and that when engaging with a media message, it is important to consider what techniques are being employed to attract buyer attention. • Understand that different people experience the same media message differently, and consider their own impressions and how others might view the message differently. • Understand that media have embedded values and points of view, and consider what values and points of views are represented. • Understand that most media messages are designed to gain profit or power, and consider why a message was sent and how an effective marketing message is created. • Select and use multiple forms of media (visual and textual) to convey information about a product and persuade a targeted audience to buy it.
3	STEM Research Notebook prompts	Individual	Formative	• Understand that all media messages are constructed and that, when engaging with a media message, it is important to consider who created it and for what purpose. • Understand that media messages are constructed using a creative language with its own rules and that when engaging with a media message, it is important to consider what techniques are being employed to attract buyer attention.

MODULE TIMELINE

Tables 3.6–3.10 (pp. 39–43) provide lesson timelines for each week of the module. These timelines are provided for general guidance only and are based on class times of approximately 45 minutes.

Table 3.6. STEM Road Map Module Schedule for Week One

Day 1	Day 2	Day 3	Day 4	Day 5
Lesson 1: The Product • Launch the module by giving students their product challenge and having them complete the Where's My Stuff? exploration (all content areas).	*Lesson 1: The Product* • In science, students begin to explore the engineering design process (EDP) by learning about chip manufacturing processes. • In math, they explore the number of broken chips in a bag. • In ELA, they begin to explore problem-solution text structures. • In social studies, they explore the effects of manufacturing, food deserts, and food swamps on communities.	*Lesson 1: The Product* • Students extend their understanding of the EDP by exploring the Doritos Locos Tacos from prototype to production. • In ELA, they use what they learned about problem structures to analyze an article about the EDP for the Doritos Locos Tacos. • In math, they continue to explore tortilla chips by finding the mean number of broken chips in a bag. • In ELA, they continue to explore problem solution, employing graphic organizers to illuminate an author's use of structure. • In social studies, students discuss banning junk food in schools.	*Lesson 1: The Product* • Students use the EDP to develop a prototype of a bag or container that will result in fewer broken chips. • In math, they continue to explore mean in the context of broken chips. • In ELA, they further develop strategies for recognizing problem-solution structure and using it to comprehend an informational article. • In social studies, students create multimedia presentations about a ban on junk food in schools.	*Lesson 1: The Product* • Students continue to develop, test, and redesign their chip packaging. • In math, students develop a survey that will allow them to determine what most people at their school consider an acceptable number of broken chips. • In ELA, students develop media literacy strategies for critically analyzing media marketing messages. • In social studies, students finalize and present their multimedia presentations.

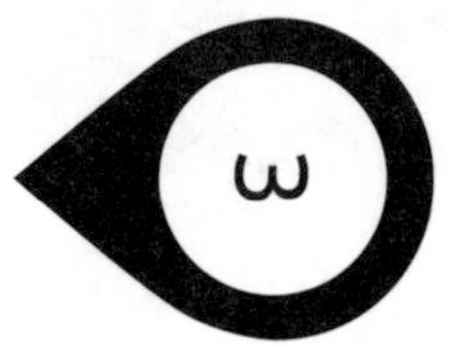

Table 3.7. STEM Road Map Module Schedule for Week Two

Day 6	Day 7	Day 8	Day 9	Day 10
Lesson 1: The Product • In science, students continue to use the EDP to test and complete their packaging. • In math, students use their data to develop a representation and choose how to answer the question about what is an appropriate number of broken chips in a bag. • They share the findings of their investigation by writing a letter on their findings to the company. • In ELA, students focus on how marketers use audience awareness (demographics) to construct effective messages. • In social studies, students continue to explore the economic impact of the Doritos Locos Tacos and the U.S. obsession with junk food.	*Lesson 1: The Product* • In science, students conduct their crush test and a winner is chosen for the Save the Chips Competition. • Students explore the caloric impact of eating an entire bag of tortilla chips. • In ELA, students work in pairs to construct an effective media message for a particular audience. • In social studies, they explore the impact of trans fats and monosodium glutamate (MSG) on their health.	*Lesson 2: The Packaging* • Students explore marketing and packaging through a variety of lenses. • In science, they look at sustainability of packaging. • In math, they explore the shapes of packages. • In ELA, students critically analyze media messages for techniques authors use to market to youth and explore Q&A text structure. • In social studies, they explore the geography of marketing.	*Lesson 2: The Packaging* • Students create a Life Cycle poster for their product's packaging and brainstorm sustainability options. • They continue to explore three-dimensional shapes in relation to packaging. • In ELA, students use Q&A text structure to support critical comprehension of an informational article about targeting in media messages. • In social studies, students examine demographics.	*Lesson 2: The Packaging* • Students continue to discuss sustainability of packaging by creating a PowerPoint presentation as a group. • In math, students explore packaging by looking at surface area and volume. • In ELA, students analyze targeted marketing in television commercials, then examine the construction and use of argumentative texts. • In social studies, students trace the journey of their product.

Table 3.8. STEM Road Map Module Schedule for Week Three

Day 11	Day 12	Day 13	Day 14	Day 15
Lesson 2: The Packaging • In science, students present their PowerPoint slides and begin working on their Life Cycle posters. • In math, students explore surface area and volume of packages. • In ELA, students use the reading strategies of summarizing and synthesizing to provide a framework for writing an argumentative response to the article. • In social studies, they explore social justice and the environment.	*Lesson 2: The Packaging* • Students complete their Life Cycle posters and present them to the class. • In math, they continue their exploration of surface area and volume. • In ELA, students use the framework from day 11 to craft a written argumentative response. • In social studies, they continue to explore social justice issues associated with their product.	*Lesson 2: The Packaging* • Students finish presenting their posters to the class. • They then brainstorm ideas about materials for their module project. • In math, they explore materials and create a scale drawing of their prototype package. • In ELA, students use media literacy concepts as an analytical framework for critically examining advertising media. • In social studies, they give their social justice presentations.	*Lesson 3: Marketing Your Product in a Global Economy* • Students explore misconceptions about colds and other viruses and how this relates to marketing of products. • In math, they build on that by learning about the Federal Trade Commission and the consequences of making false advertising claims. • In ELA, students apply media literacy concepts as a framework to construct an effective marketing message about their product. • In social studies, they explore the marketing strategies of Coca-Cola.	*Lesson 3: Marketing Your Product in a Global Economy* • In science, students continue to explore viruses and how misconceptions about them are used in marketing campaigns. • In math, they look at misleading statistics. • In ELA, students use nonfiction reading strategies and collaborative communication to learn about marketing techniques, ultimately designing a marketing campaign for their product. • In social studies, they discuss the difference between needs and wants.

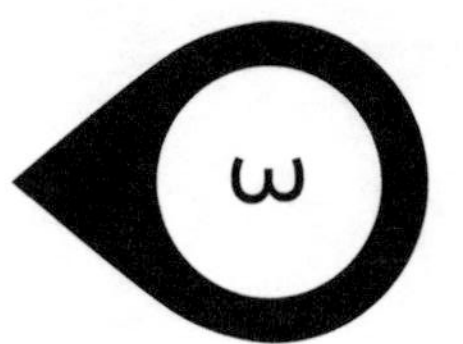

Table 3.9. STEM Road Map Module Schedule for Week Four

Day 16	Day 17	Day 18	Day 19	Day 20
Lesson 3: Marketing Your Product in a Global Economy • Students explore the nature of science. • In math, they discuss how to protect themselves from misleading statistics. • In ELA, students continue working on marketing plans. • In social studies, they explore the impacts of chocolate consumption on one community in Ghana.	*Lesson 3: Marketing Your Product in a Global Economy* • Students explore the role of media in distributing ideas about science. • Students develop a social media campaign to address some misconceptions about science. • In math, students explore marketing metrics and problem solving. • In ELA, students continue to work on marketing plans. • In social studies, they explore the life cycle of their favorite chocolate bar.	*Lesson 3: Marketing Your Product in a Global Economy* • Students give their science presentations to the class. • In math, they continue to explore marketing metrics such as return on investment. • In ELA, students learn about logo design. • In social studies, they talk about economic interdependence.	*Lesson 3: Marketing Your Product in a Global Economy* • In science and math, students work on their module projects. • In ELA, students create effective logos for their products. • In social studies, they present their findings to the class and discuss interdependence in depth.	*Lesson 3: Marketing Your Product in a Global Economy* • In science and math, students work on their module projects. • In ELA, they work on all components of the advertising campaign for the product. • In social studies, they explore a case study to determine the impact of an economic downturn on the global economy.

Table 3.10. STEM Road Map Module Schedule for Week Five

Day 21	Day 22	Day 23	Day 24	Day 25
Lesson 3: Marketing Your Product in a Global Economy • In science, math, and ELA, students work on their module projects and advertising campaigns. • In social studies, students continue their discussion on economic interdependence.	*Lesson 3: Marketing Your Product in a Global Economy* • In science, math, and ELA, students work on their module projects and advertising campaigns. • In social studies, students explore the impact the U.S. economy has on the rest of the world.	*Lesson 3: Marketing Your Product in a Global Economy* • Students participate in the competition.	*Lesson 3: Marketing Your Product in a Global Economy* • Students participate in the competition.	*Lesson 3: Marketing Your Product in a Global Economy* • Students answer questions about what they learned in the module.

RESOURCES

The media specialist can help teachers locate resources for students to view and read about packaging and related content. Special educators and reading specialists can help find supplemental sources for students needing extra support in reading and writing. Additional resources may be found online. Community resources for this module may include marketing specialists, packaging experts, economists, and bloggers.

REFERENCES

Johnson, C. C., T. J. Moore, J. Utley, J. Breiner, S. R. Burton, E. E. Peters-Burton, J. Walton, and C. L. Parton. 2015. The STEM Road Map for grades 6–8. In *STEM Road Map: A framework for integrated STEM education*, ed. C. C. Johnson, E. E. Peters-Burton, and T. J. Moore, 96–123. New York: Routledge. *www.routledge.com/products/9781138804234.*

Keeley, P., and R. Harrington. 2010. *Uncovering student ideas in physical science, volume 1: 45 new force and motion assessment probes.* Arlington, VA: NSTA Press.

National Research Council (NRC). 1997. *Science teaching reconsidered: A handbook.* Washington, DC: National Academies Press.

WIDA. 2012. 2012 Amplification of the English language development standards: Kindergarten–grade 12. *www.wida.us/standards/eld.aspx.*

PACKAGING DESIGN LESSON PLANS

Adrienne Redmond-Sanogo, Sue Christian Parsons, Janet B. Walton, Carla C. Johnson, Erin Peters-Burton, Juliana Utley, and Toni A. Ivey

Lesson Plan 1: The Product

This lesson provides an overview of the module and introduces the challenge. Students also build necessary background knowledge on problem-solution text structures and the engineering design process (EDP).

ESSENTIAL QUESTIONS

- What is the engineering design process, and how do engineers use it to solve problems?
- What are some ways that companies create, design, and market products to entice consumers into buying these product?
- How is mathematics used as a tool for solving design, marketing, and shipping issues?

ESTABLISHED GOALS AND OBJECTIVES

At the conclusion of this lesson, students will be able to do the following:

- Explain how companies purposely target specific audiences to maximize profits when creating, designing, and marketing products.
- Understand and calculate the mean as a measure of central tendency.
- Understand problem-solution text structures, and use that strategy to understand authentic literature.

TIME REQUIRED

- 7 days (approximately 45 minutes each day; see Tables 3.6 and 3.7, pp. 39–40)

MATERIALS

Required Materials for Lesson 1

- STEM Research Notebooks (1 per student; see p. 26 for STEM Research Notebook student handout)
- Handouts (attached at the end of this lesson)
- Computer and projector for students to watch videos and share ideas
- Computers, tablets, or laptops with internet access for student research and presentations
- Engineering design process (EDP) poster (Figure 2.1 on p. 10 can be enlarged to poster size)
- Chart paper

Additional Materials for Where's My Stuff?

- Student-provided digital cameras or smartphones to take digital images

Additional Materials for Science Connection

- Pictures of Doritos Locos Taco and Doritos Locos Taco prototype
- Problem-Solution Graphic Organizer (1 per student; attached at the end of this lesson)

Additional Materials for Save the Chips Competition

- Cardboard box (1 large box per team)
- Duct tape (1 roll per team)
- 4 gallon size plastic storage bags (1 per team)
- Craft sticks (10 per team)
- Glue
- Rubber bands (10 per team)
- Paper (3 sheets per team)
- Large bag of tortilla chips, preferably triangle-shaped chips (1 per team)
- Safety glasses or goggles

Additional Materials for Mathematics Class

- Paper towels

- Too Many Broken Chips Exploration Sheet (1 per student; attached at the end of this lesson)
- Sticky notes
- Poster board or butcher paper

Additional Materials for Message in a Bottle

- Recording of the song "Message in a Bottle" by the Police, plus a copy of the lyrics
- 5 index cards on which are written different demographic categories
- Markers
- A variety of product packages with informative labels

SAFETY NOTES

1. All students must wear safety glasses or goggles during the setup, hands-on, and take down segments of the activities.
2. Caution students not to eat any food used in this activity.
3. If any students have food allergies, you may need to create fictitious bags of chips instead of using real tortilla chips.
4. Make sure all materials are put away after completing the activity.
5. Wash hands with soap and water upon completing this activity.

CONTENT STANDARDS AND KEY VOCABULARY

Table 4.1 (p. 48) lists the content standards from the *Next Generation Science Standards (NGSS), Common Core State Standards (CCSS),* and the Framework for 21st Century Learning that this lesson addresses, and Table 4.2 (p. 50) presents the key vocabulary. Vocabulary terms are provided for both teacher and student use. Teachers may choose to introduce some or all of the terms to students.

Table 4.1. Content Standards Addressed in STEM Road Map Module Lesson 1

NEXT GENERATION SCIENCE STANDARDS

PERFORMANCE EXPECTATIONS

- MS-ETS1-1. Define the criteria and constraints of a design problem with sufficient precision to ensure a successful solution, taking into account relevant scientific principles and potential impacts on people and the natural environment that may limit possible solutions.
- MS-ETS1-2. Evaluate competing design solutions using a systematic process to determine how well they meet the criteria and constraints of the problem.
- MS-ETS1-3. Analyze data from tests to determine similarities and differences among several design solutions to identify the best characteristics of each that can be combined into a new solution to better meet the criteria for success.

DISCIPLINARY CORE IDEAS

ETS1.A. Defining and Delimiting Engineering Problems

- The more precisely a design task's criteria and constraints can be defined, the more likely it is that the designed solution will be successful. Specification of constraints includes consideration of scientific principles and other relevant knowledge that are likely to limit possible solutions.

ETS1.B. Developing Possible Solutions

- There are systematic processes for evaluating solutions with respect to how well they meet the criteria and constraints of a problem.
- Sometimes parts of different solutions can be combined to create a solution that is better than any of its predecessors.

ETS1.C. Optimizing the Design Solution

- Although one design may not perform the best across all tests, identifying the characteristics of the design that performed the best in each test can provide useful information for the redesign process—that is, some of those characteristics may be incorporated into the new design.

CROSSCUTTING CONCEPTS

Influence of Science, Engineering and Technology on Society and the Natural World

- All human activity draws on natural resources and has both short- and long-term consequences, positive as well as negative, for the health of people and the natural environment.
- The uses of technologies and any limitations on their use are driven by individual or societal needs, desires, and values; by the findings of scientific research; and by differences in such factors as climate, natural resources, and economic conditions. Thus technology use varies from region to region and over time.

Continued

Table 4.1. (*continued*)

COMMON CORE STATE STANDARDS FOR MATHEMATICS
MATHEMATICAL PRACTICES • MP1. Make sense of problems and persevere in solving them. • MP3. Construct viable arguments and critique the reasoning of others. • MP4. Model with mathematics. • MP5. Use appropriate tools strategically. • MP6. Attend to precision.
MATHEMATICAL CONTENT • 6.GA.2. Find the volume of a right rectangular prism with fractional edge lengths by packing it with unit cubes of the appropriate unit fraction edge lengths, and show that the volume is the same as would be found by multiplying the edge lengths of the prism. Apply the formulas $V = l\ w\ h$ and $V = b\ h$ to find volumes of right rectangular prisms with fractional edge lengths in the context of solving real-world and mathematical problems. • 6.GA.4. Represent three-dimensional figures using nets made up of rectangles and triangles, and use the nets to find the surface area of these figures. Apply these techniques in the context of solving real-world and mathematical problems. • 6.SP.A.1. Recognize a statistical question as one that anticipates variability in the data related to the question and accounts for it in the answers. • 6.SP.A.3. Recognize that a measure of center for a numerical data set summarizes all of its values with a single number, while a measure of variation describes how its values vary with a single number. • 6.SP.B.5.B. Describing the nature of the attribute under investigation, including how it was measured and its units of measurement.
COMMON CORE STATE STANDARDS FOR ENGLISH LANGUAGE ARTS
READING STANDARDS • RH.6–8.4. Determine the meaning of words and phrases as they are used in a text, including vocabulary specific to domains related to history/social studies. • RI.6.2. Determine a central idea of a text and how it is conveyed through particular details; provide a summary of the text distinct from personal opinions or judgments. • RI.6.6. Determine an author's point of view or purpose in a text and explain how it is conveyed in the text.
WRITING STANDARDS • W.6.4. Produce clear and coherent writing in which the development, organization, and style are appropriate to task, purpose, and audience. • W.6.6. Use technology, including the Internet, to produce and publish writing as well as to interact and collaborate with others; demonstrate sufficient command of keyboarding skills to type a minimum of three pages in a single sitting.

Continued

Table 4.1. (*continued*)

WRITING STANDARDS (*continued*)

- WHST.6–8.4. Produce clear and coherent writing in which the development, organization, and style are appropriate to task, purpose, and audience.
- WHST.6–8.6. Use technology, including the Internet, to produce and publish writing and present the relationships between information and ideas clearly and efficiently.
- WHST.6–8.8. Gather relevant information from multiple print and digital sources, using search terms effectively; assess the credibility and accuracy of each source; and quote or paraphrase the data and conclusions of others while avoiding plagiarism and following a standard format for citation.
- WHST.6.10. Write routinely over extended time frames (time for reflection and revision) and shorter time frames (a single sitting or a day or two) for a range of discipline-specific tasks, purposes, and audiences.

FRAMEWORK FOR 21ST CENTURY LEARNING

Information, Media, and Technology Skills

- Information Literacy
- Media Literacy
- ICT Literacy

Life and Career Skills

- Initiative and Self-Direction
- Social and Cross Cultural Skills

Table 4.2. Key Vocabulary in Lesson 1

Key Vocabulary	Definition
consumer	a person or organization that buys goods or services for one's own use
constraints	circumstances or resource limitations that limit the possible solutions to a problem
convergent ideation	when a team or individual considers a problem and narrows the number of solutions to just a few
cost-effective	producing good value for the amount of money spent
demographics	the statistical data about characteristics of a population, such as age, income, and education

Continued

Table 4.2. (*continued*)

Key Vocabulary	Definition
divergent ideation	when a team or individual considers a problem with the goal of coming up with many possible solutions based upon different ideas from a variety of perspectives.
food desert	a usually low-income area whose residents cannot easily access stores selling affordable, healthful foods
food swamp	an area with little to no access to affordable, healthful foods but an abundance of convenience stores and fast food restaurants
icon	a widely recognized person or symbol
ideation	the process of forming ideas or concepts
ideation session	a brainstorming meeting in which a group comes together to create and discuss new ideas
innovation	introduction of something new; a new idea, method, or product
junk food	food that is low in nutritional value but high in calories, often highly processed and needing little or no preparation, such as packaged snacks and many fast foods
logo	a symbol or design used by a company or organization to identify its brand or products
manufacturer	a person or business that owns or runs a plant that produces goods
marketing	the act or process of advertising, selling, and distributing goods or services from producers to consumers
mean (arithmetic)	the average of several values, calculated by adding the values together and then dividing the total by the number of values
media literacy	the ability to read and analyze the many media messages around us with an awareness of the goals of these messages and how they may affect people
median (arithmetic)	the middle number in a set of numbers; if the set has an even number of values, the average of the two middle numbers
mode (arithmetic)	the number that occurs most often in a set of numbers
monosodium glutamate (MSG)	a white crystalline sodium salt derived from an amino acid called glutamic acid, used as a flavor enhancer in foods

Continued

Table 4.2. (*continued*)

Key Vocabulary	Definition
prototype	an original, first model of a new product, from which other forms are developed or copied
refute	to prove something to be wrong, false, or erroneous
researchable	able to be answered using a systematic inquiry or investigation
shipping	the transport of goods; in this module, the transport of products from the manufacturer to the consumer
symbol	something used to represent something else; in this module, a material object used to represent an abstract concept or idea
trans fat	also called trans fatty acid, a type of solid fat made from vegetable oil through a chemical process called hydrogenation; used in many processed foods but considered unhealthy because an excess of trans fats in the diet can raise cholesterol levels and increase the risk of heart disease
visual imagery	pictures or graphics designed to appeal to the consumer through the sense of sight

TEACHER BACKGROUND INFORMATION

This lesson introduces the module and the final challenge by engaging students with packaging by asking them to respond to a display of products and consider the appeal of the packaging to buyers. Students will receive a copy of the module challenge guidelines as an introduction to the module's content. The lesson also provides students with the opportunity to learn about the engineering design process, engineer solutions to real problems, and use that knowledge to think about social issues. The information in this section will help you engage your students in this lesson.

Engineering Design Process

The engineering design process (EDP) is a series of steps that engineers go through to solve problems. First, the engineer or engineering team defines the problem. Next, the team does research into the problem and brainstorms potential solutions. Once the team has come up with a solution, it then develops a prototype/program and tests the solution. Finally, the team redesigns as necessary or moves ahead to present the solution. Redesign after a design failure is an important component of the engineering design process. A graphic image of the EDP is attached at the end of this lesson.

For more background on the EDP, teachers can visit the following websites:

- *www.sciencebuddies.org/engineering-design-process/engineering-design-process-steps.shtml*
- *www.eie.org/overview/engineering-design-process*
- *www.teachengineering.org/engrdesignprocess.php*
- *https://curriculum.vexrobotics.com/curriculum/intro-to-engineering/what-is-the-engineering-design-process.html*

Ideation Sessions

One useful brainstorming technique is an ideation session, in which individuals come together as a group to create and discuss new ideas. Ideation sessions are used frequently in commercial product design. Ohler and Samuel (2013) list five important elements of successful ideation sessions:

- Define the opportunity clearly
- Allow for separate phases of divergent and convergent thinking
- Stimulate thinking inside and outside the box
- Include all problem solving styles
- Ensure an efficient facilitation process for ideation teams

Divergent ideation occurs when a team or individual considers a problem with the goal of coming up with many possible solutions based on different ideas from a variety of perspectives. *Convergent ideation* can follow divergent ideation and occurs when a team or an individual considers a problem and narrows the number of solutions to just a few. After students have posed all plausible possibilities using divergent ideation, they will need to engage in convergent ideation to reduce the number of ideas to a set they can test. In preparation for this task, the class will establish ground rules for "killing" ideas and "pursuing" ideas. This could be as simple as assigning all ideas a number or letter and having team members vote for the three ideas they think are the best, tallying votes and killing all but the three or four ideas that received the most votes. For other suggestions on how to converge group thinking, see *www.interaction-design.org/literature/article/how-to-select-the-best-idea-by-the-end-of-an-ideation-session*.

Symbols and Iconic Products

Symbolism is the use of something to represent something else. In this module, a *symbol* is a material object used to represent an abstract idea. *Iconic* products are those that are instantly recognized by the consumer and widely regarded as the exemplar of that type of product. An iconic product may, in the mind of the consumer, become almost synonymous with the product in general; for instance, a sneezer frequently asks for a Kleenex instead of a tissue, regardless of what brand name is on the package.

Three-Act Lessons

Three-act lessons are a method of teaching mathematics that integrates technology, mathematics, and English language arts (ELA), based on the three-act tasks created by Dan Meyer, who discusses them in his blog at *http://blog.mrmeyer.com/2013/teaching-with-three-act-tasks-act-one/*. The first act is a video that engages students in a scenario. It is very visual and contains few or no words. Its purpose is to create dissonance and have students begin to think of questions they want to answer in the lesson. After students watch the video, they are asked what they wonder and what they noticed. Students are then asked to form questions about what they saw. The idea is to increase students' curiosity about what was happening in the video. You will have an idea in mind for what you want the students to accomplish and how they will solve that problem. However, students will ask other questions that you can also pursue.

In the second act, students are asked what they need to solve the problem and then are allowed them to solve the problem. The teacher may provide information as needed throughout this section. Finally, the third act contains the great reveal, when the results of student work are shown.

Career Connections

There are many different careers associated with the development, manufacturing, shipping, and display of snack products. As career connections related to this lesson, you may wish to introduce the following:

- *Chemical Engineer:* Chemical engineers consider how the chemistry of the food will interact with the packaging. For more information, see *https://www.bls.gov/ooh/architecture-and-engineering/chemical-engineers.htm.*
- *Mechanical Engineer:* Mechanical engineers develop and modify machines to create the products. For more information, see *https://www.bls.gov/ooh/architecture-and-engineering/mechanical-engineers.htm.*
- *Physicist and Mathematician:* Physicists and mathematicians explore size, shape, mass, and structure of the products and manufacturing equipment. For more information, see *https://www.bls.gov/ooh/life-physical-and-social-science/physicists-and-astronomers.htm* and *https://www.bls.gov/ooh/math/mathematicians-and-statisticians.htm.*

Media Literacy

Media literacy refers to the ability to read and analyze the many media messages around us with an awareness of the goals of these messages and how they may affect people. According to the Center for Media Literacy (CML),

> *Media Literacy is a 21st century approach to education. It provides a framework to access, analyze, evaluate, create and participate with messages in a variety of forms—from print to video to the Internet. Media literacy builds an understanding of the role of media in society as well as essential skills of inquiry and self-expression necessary for citizens of a democracy. (CML 2015)*

A central goal of instruction in this module is for learners to become critical readers and producers of multimedia messages, particularly marketing messages. These lessons focus on developing understanding of media literacy concepts and, with these concepts in mind, learning to approach texts from a critical inquiry stance. The five core media literacy concepts and five related key questions identified by the CML are woven through the lessons. Review these concepts and questions, as well as the additional information, at *www.medialit.org/reading-room/five-key-questions-form-foundation-media-inquiry*. Another helpful source is "Media Literacy in the Middle School" at *www.medialit.org/reading-room/media-literacy-middle-school.*

Nonfiction Text Structures and Features

Throughout this unit, students read and write nonfiction informational texts from a variety of authentic sources and for a variety of purposes. Although nonfiction informational texts constitute the majority of texts read in secondary schools, in the workplace, and in day-to-day interactions, much early reading instruction tends to focus on reading fiction. Because nonfiction texts are structured differently and serve different purposes than fictional narrative texts, readers need specific instruction on how to read and compose them. Nonfiction text structures specifically addressed in these lessons are question and answer and problem and solution. Other common nonfiction text structures include cause and effect, comparison and contrast, description, and sequence. Common nonfiction text features include fonts and special effects (e.g., titles, headings, boldface print, italics, and bullets), textual cues such as "for example," illustrations and photographs, graphics, and text organizers (e.g., table of contents, glossary, and index.) Teaching students to recognize and use these structures in texts supports their abilities to understand and compose nonfiction texts.

For additional information and suggestions for teaching, see *Nonfiction Matters: Reading, Writing, and Research in Grades 3–8,* by Stephanie Harvey. For nonfiction text structures to explore in the *New York Times,* see "Compare-Contrast, Cause-Effect,

Problem-Solution: Common Text Types in the Times," by Katherine Schulten, on using the *Times* to teach text type (*http://learning.blogs.nytimes.com/2011/12/12/compare-contrast-cause-effect-problem-solution-common-text-types-in-the-times*).

For specific enrichment related to reading and writing nonfiction, explore the following:

- "Teaching Reading 3–5: Summarizing Nonfiction" (adaptable to upper grades): *www.learner.org/workshops/teachreading35/classrooms/cv8.html*
- "Organizing Ideas From Multiple Sources": *www.learner.org/courses/readwrite/video-detail/organizing-ideas-multiple-sources.html*
- "Teaching Content Through Literacy": *www.learner.org/courses/readwrite/video-detail/teaching-content-through-literacy.html*
- "Writing Across the Curriculum": *www.learner.org/workshops/writing35/session8/index.html*

Reading Comprehension and Writing Strategies

The reading comprehension strategies specifically addressed in this module include determining important information, summarizing and synthesizing, and questioning—all strategies that are vital for reading the kinds of texts students will encounter across subject areas. See *Strategies That Work: Teaching Comprehension for Understanding and Engagement,* by Stephanie Harvey and Anne Goudvis, for more information about these and other vital strategies for middle school readers. Also see the National Writing Project's "Resource Topics: Teaching Writing" at *www.nwp.org/cs/public/print/resource_topic/teaching_writing* for a topical directory.

Students in this module write for a variety of purposes and audiences, so genre-specific writing strategy instruction is included as well. Particular attention is given to developing argumentative/persuasive texts as part of the larger module focus on media and marketing. While some specific instructional strategies are suggested, teachers should also be aware of opportunities to support other content-area teachers as they work with various writing assignments. In addition, teachers should remain cognizant of learner background regarding navigating the internet safely and productively, teaching and supporting those skills as needed.

Flow Charts

The use of flow charts has been shown to be highly effective in supporting diverse learners across content areas. In this lesson, flow charts help students visualize the problem-solution structures embedded in texts. The flow chart is introduced as a way to graphically represent the structure of a problem-solution article. As students continue in the

unit, they will use flow charts to support their own problem-solution text construction. For more information on using flow charts with middle school learners, consult *Inclusion Strategies for Secondary Classrooms: Keys for Struggling Learners,* by M. C. Gore.

Picture Books

Picture books, often considered material for lower elementary learners, are powerful resources for upper-level learners as well. Because of their short length, the underlying structures of the texts are more readily accessible, allowing learners to focus on understanding how the texts are constructed. This knowledge provides the foundation for understanding longer texts. In addition, picture books allow for easy exploration of multiple topics, perspectives, and genres in a short time. Finally, because of the wide range of complexity and readability, picture books readily allow for differentiation among learners.

Food Deserts and Food Swamps

Following are some websites for additional information:

- *http://mic.com/articles/7176/obesity-food-deserts-have-given-way-to-food-swamps*
- *http://voices.washingtonpost.com/all-we-can-eat/food-politics/food-deserts-vs-swamps-the-usd.html*
- *http://brownisthenewpink.com/2014/05/27/food-deserts-and-swamps-social-justice-issue*

COMMON MISCONCEPTIONS

Students will have various types of prior knowledge about the concepts introduced in this lesson. Table 4.3 (p. 58) outlines some common misconceptions students may have concerning these concepts. Because of the breadth of students' experiences, it is not possible to anticipate every misconception that students may bring as they approach this lesson. Incorrect or inaccurate prior understanding of concepts can influence student learning in the future, however, so it is important to be alert to misconceptions such as those presented in the table.

Table 4.3. Common Misconceptions About the Concepts in Lesson 1

Topic	Student Misconception	Explanation
Engineering design process (EDP)	Engineers use only the scientific process to solve problems in their work.	Science processes are used to test predictions and explanations about the world. The EDP, on the other hand, is used to create a solution to a problem. In reality, engineers use both processes.
Measures of center	The average of a data set is the value that lies in the middle of the data when it is ordered.	The term *average* can be confusing to students because it can refer to three different measures of central tendency—mean, median, and mode. These terms should be used explicitly in statistical discussions. The *mean* is calculated by summing all quantities in the data set and dividing this sum by the number of quantities; the *median* is found by ordering values sequentially and identifying the value that lies in the middle of the data set; the *mode* is the value most frequently found in a data set.
	The mean of any data set is meaningful; students may attempt to calculate a mean of frequency counts applied to categorical data sets.	Categorical data represent groups of things, such as gender, favorite vacation spots, or age group. Whereas frequency counts can be conducted for these data, means calculated from the data are not meaningful. For example, a poll of favorite vacation spots indicates that 7 students prefer the mountains, 12 the beach, 10 amusement parks, and 4 urban destinations. A student may calculate that the mean of these data as (7 + 12 + 10 + 4) ÷ 4 = 8.25. This value lacks meaning, however, since it is not possible to indicate the mean of the categories (i.e., the beach, urban destinations).

PREPARATION FOR LESSON 1

Review the Teacher Background Information (p. 52), assemble the materials for the lesson, make copies of student handouts, and preview the recommended videos and websites. Have your students set up their STEM Research Notebooks (see pp. 25–26 for discussion and student instruction handout). Students should include all work for the module in the STEM Research Notebook, so you may wish to have them include section dividers in their notebooks.

A week before you plan to begin teaching this lesson, have students complete the Where's My Stuff? activity. A few days before the Too Many Chips activity in mathematics, tell students that they will each need to bring a bag of tortilla chips (triangle-shaped chips). Students may provide a range of bag sizes. You may choose to provide all the chips if this might pose problems for economically disadvantaged students.

For the science connection, choose a video about how chips are manufactured (some examples are listed on p. 61). Print out copies of "Deep Inside Taco Bell's Doritos Locos Taco," by Austin Carr, at *www.fastcompany.com/3008346/deep-inside-taco-bells-doritos-locos-taco.* Read the article ahead of time so you can be prepared to discuss with students the EDP and problem-solution text structure addressed in this piece. Prepare a poster-sized copy of the EDP graphic (p. 79) to share with students. Make a chart that shows appropriate conversations to have during a convergent ideation session. Prepare a testing station for the Save the Chips competition. This station should include a space where packages can be dropped from a 5 foot height and thrown 5 feet, and it should include a 5-pound weight that can be placed on teams' package designs.

For ELA class, select a variety of articles with a problem-solution focus that are appropriate for the learners in your classroom. One good source for such articles is the *New York Times* "Fixes" blog at *http://opinionator.blogs.nytimes.com/category/fixes/. Time for Kids* is a good source for more easily accessible texts if needed to support learners. Also search online for cartoons or other images that show people stepping on chewing gum. Look for ones that will easily elicit an "Eew!" response from middle school learners. Gather a variety of flow-chart configurations that would work well to show a problem-solving process. You can find these flow charts by conducting an Internet image search using search terms such as *problem solving flow chart examples*. The following article from Sciencing.com, "How to Solve Math Problems Using a Flow Chart" by Shelley Frost, may also be useful: *https://sciencing.com/solve-math-problems-using-flowchart-7840920.html*.

Select a problem-solution picture book for modeling from the list suggested on pages 66–67. Choose commercials for students to analyze. One helpful resource is iSpot.tv, a website that tracks advertising media. The page for "Toys & Games TV Commercials" provides numerous links to these ads: *www.ispot.tv/browse/w.dQ/life-and-entertainment/toys-and-games.* Optional: You may wish to prepare a minilesson on nonfiction features if any students do not understand nonfiction text.

For social studies, print out copies of the "Why This Food Is Garbage!" article at *http://whythisfoodisgarbage.blogspot.com/2011/01/doritos.html* to hand out to students. Have Google Maps ready to use on a computer.

LEARNING COMPONENTS

Introductory Activity/Engagement

Connection to the Challenge: After launching the module (see Module Launch, p. 27), begin each day of this lesson by directing students' attention to the essential questions and the module challenge. It may be helpful to remind them to revisit the Product Design Challenge Guidelines handout (pp. 93–94) each day.

Where's My Stuff?

Have students complete this activity a week before you plan to begin teaching Lesson 1. Introduce the STEM Research Notebooks and give students the following instructions:

- Plan a trip to a grocery store with three to five other people of different ages and backgrounds.
- Have each person select one item he or she wants from the store. Take pictures of each item, showing the item's location and the items around it.
- In your STEM Research Notebook, record the ages and genders of all the people who went with you.
- Answer the following questions: Where did you find each of the items in the store? What is the store's assumption about you and the people who shopped with you?
- Create a graph, infographic, or map of where you and the people who shopped with you found their items.

Have students share their observations and visual artifacts with their classmates, and hold a class discussion on the placement of products.

STEM Research Notebook Prompt

Provide students with the Product Design Challenge Guidelines Handout and the Product Design Challenge Rubric (Rubric 1, p. 85). Have them paste or staple these into their STEM Research Notebooks. Pose the following question again, and have them respond in their STEM Research Notebooks: *What aspects of the products displayed might make them appealing to a buyer?*

Activity/Exploration

Science Connection: Show students a video on how tortilla chips are manufactured and packaged, such as the YouTube video "How It's Made—Tortilla Chips" at *https://www.youtube.com/watch?v=OFh_q4e_boI* or "How Tortilla Chips are Made" at *www.youtube.com/watch?v=QzIdZGOR9vo.*

STEM Research Notebook Prompt

After students have watched the video, have them respond to the following questions in their STEM Research Notebooks, and then discuss their responses using a think-pair-share strategy.

- *What were some things you noticed about the way the chips were made, packaged, and shipped?*
- *Why did the manufacturer choose to shape the chips that way?*
- *How were the chips manufactured?*
- *What problems could arise that engineers would have to address in the manufacturing and shipping processes?*
- *What had to be created and designed to make this process work?*

Explain to students that engineers had to think through the manufacturing and packaging process. They had to design machines and refine the process to make it safer, more efficient, and cost-effective. Let students know that they will be figuring out how to package these chips.

Introduce the EDP. Review each of the elements with students, and remind students that they will document their use of the EDP in their STEM Research Notebooks throughout this module. Students should create a section for each step of the EDP and make notes about their work for each step, including labeled sketches and appropriate citations for any information they identify from research. Provide students with the EDP graphic (p. 79), and have them place a copy in their STEM Research Notebooks.

Discuss the different components of the EDP in the context of the video the students watched. Have students label a page in their STEM Research Notebooks as DEFINE. Ask students to respond to the following questions under the DEFINE heading, and then discuss as a class.

- What were some of the problems that the chip manufacturers might have encountered when trying to make sure the tortillas have the perfect crunch?

- What do you think were the needs and constraints that they had to deal with through the process?

Underneath the students' responses to the DEFINE questions, have them write the heading LEARN. Have students research possible issues and problems that manufacturers faced when designing, manufacturing, packaging, and shipping chips. They should write their research findings in their STEM Research Notebooks. Then, have students share their findings with the class.

Beneath the LEARN responses, have the students write the heading PLAN. Divide students into groups of two or three for the module challenge, and have them discuss as a team how the manufacturer solved those problems and brainstorm other possible solutions to these problems. Tell students to write their ideas in their STEM Research Notebooks, then have them share with the rest of the class. Discuss with students that the *try*, *test*, and *decide* phases of the EDP would be used to create and build a prototype, test and evaluate the prototype, and improve and redesign as needed.

The following day, ask students if they have ever eaten a Doritos Locos Taco. Share images of these tacos as well as the prototype of the Doritos Locos Taco. Task students with brainstorming about the EDP they learned about previously. Have students list the different phases of the EDP, and discuss what happens in each of these phases.

STEM Research Notebook Prompt

Ask students to respond to the following questions in their STEM Research Notebooks, and then discuss using a think-pair-share strategy.

- *What process do you think Taco Bell had to go through to create the Doritos Locos Taco?*
- *What do you think were some problems that the company had to address along the way?*

Write students' ideas on the board or on a chart. Tell students that they are going to read "Deep Inside Taco Bell's Doritos Locos Taco: From Handshake Deals to Experiments at Home Depot, the History of Taco Bell's Disruptive Faux Cheese-Dusted Taco," by Austin Carr. As students read, have them answer the following questions in the margins of the article:

- What is interesting?
- What do you wonder about?
- What is confusing?

Tell students to circle any words that they are unsure about so that they can discuss the meanings with their group members. Have students turn and talk to their group members about the article and then share out loud as a class.

Provide students with copies of the Problem-Solution Graphic Organizer to use to help them think about problem solution. Revisit the text to identify problems encountered and solutions devised in the process of designing, manufacturing, shipping, and marketing the Doritos Locos Tacos. Remind students that they have been using this process in their ELA classes. Come back as a group to discuss their problem-solution organizers.

Save the Chips Competition

Pose the following problem: "You have received letters from numerous customers complaining that your bags of chips have too many broken chips inside them." Ask students to define the problem and write this in their STEM Research Notebooks. Let students know they will be using the EDP to create a solution to this problem. Tell students that each group should design a container that will protect the chips from being broken when dropped from 5 feet, when thrown 5 feet, and when a 5-pound weight is placed on top of the container.

Provide students with the materials listed for this competition on page 46. Tell students that they can use any or all of these materials to create a prototype of their design using the EDP. The package of chips must fit inside the package they create, and the goal is for the designed package to withstand being dropped 5 feet, being thrown 5 feet, and having a 5-pound weight placed on it without breaking and without damaging the chips inside it. Ask students to identify the constraints they will face in this challenge (e.g., can only use materials provided, chip bag must fit inside packaging, limited time). Students should label a page for each step of the EDP in their STEM Research Notebooks (i.e., *Define, Learn, Plan, Try, Test, Decide*). Working in teams, students should share their ideas about each step and record team ideas in their notebooks.

Remind students that in the plan phase they should create a labeled diagram of their design in their notebooks. During the test phase of the EDP, student teams can test their designs at the testing station you prepared (see Preparation for Lesson 1, p. 59). Students should record the results of the test and possible improvements they could make to their design on the Decide page of their notebooks. After testing their designs, teams should revise their designs. After all teams have tested and revised their designs, teams will compete against one another to determine which packages best withstand testing (see Elaboration/Application of Knowledge, p. 72).

Mathematics Class: Students will complete the activities below.

Three-Act Lesson

Show students the "Tortilla Chips Act One" video at *www.youtube.com/watch?v=TbO79YIBu00*. The video shows a young man holding a bag of chips in each hand, one small and one large. He looks inside the small bag, and then pulls out a chip and eats it. Next, he opens the large bag, looks inside, and then eats a chip from this bag.

Ask students what they notice about the video. Then, ask them what they wonder about the video or what questions they have. Ask the class the following questions:

- Why is one bag so much bigger than the other? Does that mean there are more chips in the larger bag?
- What do you think is an estimate of the number of tortilla chips that can fit in the small bag? The large bag?
- What is an estimate that would be too big?
- What is an estimate that would be too small?

Give students the Act-Two Images handout (pp. 82–83). The images in this handout show the outside and inside of the large and small packages, the nutritional information and serving size labeling, and the two bags of tortilla chips poured out onto paper towels. Ask students to consider how they would formulate an answer to the question, "How many chips are in each bag?"

Have students work in groups of two or three to devise ways to answer this question. Then, have students share some of their solution strategies with the class. (As students are solving the problem, take notice of students' solutions, and be prepared to have students share their findings moving from less sophisticated to more sophisticated solutions.) After you have students share a variety of solution strategies, show connections among the different solution strategies.

Near the end of the class, give students the Act-Three Images handout (p. 84). The images in this handout show the results of counting all the chips in the bags. Point out that there are lots of broken chips in the bag.

Ask students, "What do we count as a full chip? What do we count as a broken chip?" Ask students to each bring in a bag of tortilla chips (triangularly shaped chips) for the next class to explore these questions further. The bag can be any size, as you want to wind up with a variety of sizes. (*Note:* Tell students a few days before this portion of the lesson that they will each need to bring a bag of tortilla chips. You may choose to provide all the tortilla chips if this might pose problems for economically disadvantaged students.)

Too Many Broken Chips

The following day, group students by the size of their chip bags. As an introduction to the activity, engage students in a discussion about the three-act lesson. Discuss as a group the following question: "What problem did we run into when we were trying to determine the number of chips in the bags?"

Have students write the heading DEFINE in their STEM Research Notebooks. Explain to students that they are going to explore how to find the percentage of broken chips in each bag. Have students work in teams to create a problem statement and record this on the DEFINE page of their notebooks.

For the exploration, students need to decide on a common definition of "broken chip." Have students record this information in their STEM Research Notebooks on a page with the heading LEARN.

Teams will also need to create a graphic that shows various levels of brokenness and how to classify each chip. Have students write the heading PLAN on a page of their STEM Research Notebooks. As a class, brainstorm some ways for students to determine how to make a complete chip from broken chips and have students record their ideas in their STEM Research Notebooks.

Next, have students empty the contents of their chip bags onto paper towels. Ask students to examine the contents of their chip bags to answer the questions on the Too Many Broken Chips Exploration Sheet (this represents the Try phase of the EDP).

Next, bring students together and have them discuss their findings and how they computed the average. You may need to conceptually develop the method for finding the average by providing students with connecting cubes and having them create a bar graph of the data. Then, have students level the bars. Ask students to try to write how this would look if they were to use number sentences. Ask students how they could write a formula for finding the arithmetic mean (average) for any set of numbers. Provide students with the following method for finding the arithmetic mean if they do not develop it:

$$arithmetic\ mean = \frac{a_1 + a_2 + a_3 + \cdots + a_n}{n}$$

Did they find the average percentage? Did they find the average number using the total? Clear up any misconceptions the students may have about the average number of broken chips.

STEM Research Notebook Prompt

At the conclusion of the lesson, students should respond to the following questions in their STEM Research Notebooks:

- *What are some things you are still wondering about the broken chips?*
- *What are some things you are still wondering about finding the mean (average) of a set of data?*
- *If you were required to teach a third grader how to find the average number of broken chips, what would you tell him or her?*
- *How could we find the average number of broken chips as a class? What would this information tell us?*

ELA Class: Have the students complete the activity below.

So You Think You've Got Problems

This activity teaches students to recognize problems and solutions in real-world texts. Show students the images you found of people stepping on chewing gum, and briefly discuss this common and annoying (gross!) problem. Every single day, human beings in every walk of life—at home, at work, and at school—encounter problems to be solved. Fortunately, we humans are really good at figuring out solutions to the problems that get in our way.

Share the short article on "Nonstick Chewing Gum," about the invention of non-stick, dissolving chewing gum to solve the problem of discarded chewing gum and the resulting sticky situations, found at *http://sciencenetlinks.com/science-news/science-updates/nonstick-chewing-gum*. Guide students in identifying and articulating the problem and solution that the article addresses. Explain that solving problems requires us to use what we know but also to consider possibilities that are different from our usual approaches. If what we are already doing is working, we wouldn't have the problem! Innovation is the process of coming up with a new way of doing something—a new idea, a new product, a new method. We must consider possible new solutions. Today, we are going to take a look at how some innovations resulted from knowledgeable, creative thinkers tackling a problem.

Read aloud a picture book that highlights the process of problem solving through innovation, including the problems that innovators encounter on the way to pursuing their great ideas. You can choose from a variety of high-quality picture books about innovative problem solving that will appeal to middle grade readers. Following are some suggested texts that fit well with this module:

- *Balloons Over Broadway: The True Story of the Puppeteer of Macy's Parade,* by Melissa Sweet
- *The Day-Glo Brothers: The True Story of Bob and Joe Switzer's Bright Ideas and Brand-New Colors,* by Chris Barton

- *Energy Island: How One Community Harnessed the Wind and Changed Their World,* by Allan Drummond
- *Mr. Ferris and His Wheel,* by Kathryn Gibbs Davis
- *Electric Ben: The Amazing Life and Times of Benjamin Franklin,* by Robert Byrd (especially the section on inventions)

As you read, encourage students to listen for problems encountered and solutions that the innovators devised. After the reading, record students' insights about problems and solutions in the text on a chart or other display. Explore with students a variety of flow charts that they can use as models, and discuss how one or two might be used to show the processes in the book you read together Then, use student input to model the creation of a flow chart, tracking the path of problem and solution described in the book as it leads to the creation of an innovative product or process. Explain that the resulting chart is a graphic representation—a map of sorts—of the problem-solving process.

Lead students to think about how various structures at school (physical structures such as the building, as well as scheduling structures such as class schedules) affect how successfully we get through the day. For example, on the first day in a new building or with a new schedule, people often get lost and confused because they aren't sure where to go or what to do. Once we know how to get around the building or understand how the schedule works, we can more easily pay attention to the important things like learning.

Introduce the concept of nonfiction text structures by explaining that good writing has a structure that helps the writer organize ideas and the reader understand the writing. Information about innovation often comes packaged in a problem-solution text structure. To understand a complex article, it helps to be able to recognize and understand the structure so you can find your way through the piece. When you understand the structure, you can more easily follow what the author is saying.

Provide students with a variety of articles that focus on real-world problems and solutions (see *New York Times* "Fixes" blog at *http://opinionator.blogs.nytimes.com/category/fixes*). Have students work collaboratively in small groups of three to five. Provide enough articles to allow for topic choice, and group students according to the articles they chose. Allow sufficient time for each student to read the article independently or with a partner (as needed or preferred).

Each group should identify, in general, the initial problem and resulting goal, steps taken to solve the problem, and the ultimate solution devised. Using one of the flow chart structures discussed earlier, or creating one of their own, students should show the problem-solution process used in their article. Then, have each group share the results of their analysis with the class. Talk together about how sharing the structure—the problem and solution process—helped everyone understand what the article was about. Remind students that problem and solution is a structure that writers use to explain these kinds

of processes. If they can recognize that structure in something they are reading, they can follow it like a map to understand the piece they are reading. If they use it to organize their own writing, they can communicate their ideas more clearly.

Social Studies Connection: Ask students how close they live to a supermarket, farmers' market, or other location where they can purchase fresh fruits and vegetables. Use Google Maps to help students find the distance. Ask students what they think about when they hear the terms *food desert* and *food swamp*. Have students each share their thoughts with a partner, and then share with the class what their partners thought these words meant. Ask students to view the map and read the short article titled "This Sobering Map Shows You All of America's Food Deserts" at *https://grist.org/food/this-sobering-map-shows-you-all-of-americas-food-deserts*. Hold a class discussion about students' experiences with food deserts. Next, pose the following question to students: "Are people who live in food swamps healthier than those who live in food deserts?" Have students share their ideas. As a class, read the article titled "Food Swamps Are the New Food Deserts" at *www.theatlantic.com/health/archive/2017/12/food-swamps/549275/* and then revisit the question about whether people who live in food swamps are healthier than those who live in food deserts.

As homework, have students explore their neighborhoods and identify sources of fresh food near them. Have students compare and contrast the availability of fresh food in different locations throughout the state and nation. Specifically, students should explore the impact that food deserts and food swamps have on them and their families, as well as on families with children outside their community.

The next day, have students discuss their findings as a class. Explain to students that there is a big push in the United States to charge a junk food tax or to ban all junk food in schools.

STEM Resource Notebook Prompt

Ask students to respond to the following prompt in their STEM Research Notebooks: *How do you feel about a ban on all junk food in schools?* Then, have a few students share their positions with the class. During the discussion, have students consider which arguments are based on fact and which ones are based on opinion.

Assign students to teams of two and have them conduct research on the internet to decide whether junk food, including chips, should be banned from the school. Ask students to find at least three important facts that support the ban of junk food and three facts that do not support the ban and to record these facts in their STEM Research Notebooks.

Explanation

Science Connection: Introduce careers associated with the production of snack products. These include chemical and mechanical engineers, physicists, and mathematicians (see Teacher Background Information section on p. 54 for more information).

Have students conduct research on the internet to determine what causes broken chips and explore how the companies protect their product from breaking. Ask students to write the heading LEARN in their STEM Research Notebooks and record what they find there. Then, have them share their research with the class. Have students write the heading PLAN beneath what they just wrote. Explain the difference between divergent and convergent ideation sessions, and tell students that they will conduct a divergent ideation session to think of a variety of solutions to the problem of broken chips. In this session, you should establish some ground rules for killing ideas and pursuing ideas. (More information on ideation sessions is provided in the Teacher Background Information section on p. 53.) Have students work in groups of two or three to brainstorm ideas. Tell them to write these ideas in their STEM Research Notebooks.

Then, bring students back together. Tell students that they will now conduct a convergent ideation session and choose one design to pursue. Instruct students to save all designs, because the first design they choose may not be effective.

STEM Resource Notebook Prompt

- *What design are you and your group going to pursue?*
- *How will you use the engineering design process to solve this problem?*
- *How will this process help you with the module challenge?*

Mathematics Class: To prepare for the lesson, read students' STEM Research Notebook prompt responses from the previous day. Take note of any misconceptions, and be prepared to support those students as they work to complete the following STEM Research Notebook prompt.

STEM Research Notebook Prompt

As students come in, have them respond to the following prompt in their Research Notebooks: *Mrs. Savannah's class did the same investigation that we did yesterday. One of the groups in her class found the following total numbers of broken chips: Student A: 22 broken chips; Student B: 34 broken chips; Student C: 25 broken chips; and Student D: 23 broken chips. Determine the mean number of broken chips in this group of students. Explain your thinking using words, pictures, and symbols.*

Have students pair up and share their solutions with their partners. As students are sharing with each other, select a few to share their processes with the class. (Try to choose students with a variety of solutions.) Discuss similarities and differences in the solution strategies, and clear up any misconceptions and errors here.

Remind students that yesterday they explored the number of broken chips in their bags and found the mean number of broken chips. They then answered some questions in their STEM Research Notebooks. Ask a few students to share their responses to the following questions:

- What are some things you are still wondering about the broken chips?
- What are some things you are still wondering about finding the mean (average) of a set of data?
- If you were required to teach a third grader how to find the mean number of broken chips, what would you tell him or her?

Have the groups each record their mean number of broken chips and their individual number of broken chips on sticky notes, and then place them on a class chart. Using the data available in the class chart, find the mean number of broken chips for the class. (Allow students to work with partners on this task.) Be sure that students can determine the number of total chips in the bag. They can use the information provided on the nutrition label to estimate the total number.

As students are exploring this task, monitor student solutions and select a few students to share in a whole-class discussion. After students have had time to find a solution, bring the whole class back together for the discussion. Pose these questions at the end of the discussion: "Do you think this number of broken chips is acceptable? Why or why not? Do you think other students in the school would feel the same way?"

ELA Class: Have students complete the activity below.

Message in a Bottle

Display an image of a message in a bottle while you play a recording of "Message in a Bottle" by the Police. Discuss the lyrics. What is this idea of a message in a bottle? Bring the discussion around to the idea that someone is sending out a message in hopes that someone will pick it up and do something in response and that the message isn't sent directly to any one person. While not as poignant, marketers do basically the same thing: they send messages about their products out into the world hoping that someone will pick up those messages and buy their products. Just like a message in a bottle, this "message on a bottle" (or can or box) must be brief yet effective, able to communicate across time and space to reach a likely responder. Refer to the Where's My Stuff? activity (see

p. 60) and discuss how the packaging influenced students' decisions about the products they wanted.

Hold a class discussion about the purposes of packaging. Have students brainstorm ideas in their groups about the goals of package design, and then have teams share their ideas with the class, creating a list of ideas. Prompt students to include purposes such as product protection during transportation of products from the manufacturer to the store, appeal to customers, providing product information, and differentiating brands from each other. Next, ask students to share ideas about the ways manufacturers might design packages to appeal to certain kinds of people. Record student ideas on a chart or other display.

Divide students into small groups of three or four, and give each group several product packages to analyze. Ask students to examine the packaging carefully for messages about intended use and appeal. They should consider the following questions:

- What are the messages the manufacturer is trying to send to the consumer regarding what the product is for and the quality and appeal of the product?
- To what sorts of consumers does the marketer seem to be aiming the message?
- How is the marketer using language (such as word choice and phrasing) and visual imagery (such as pictures or graphics) to entice and position the consumer?

Have students share and discuss their findings. Introduce the concept of demographics. Marketers need to determine what type of person would be most likely to buy a product, based on group characteristics. These broad categories describing people are called demographics. Common demographic categories include age, culture or ethnic group, education level, household composition (married or single and number of children), and professional/employment status. Discuss each category briefly to make sure students understand each one.

Explain that each characteristic has a potential effect on purchasing choices. You can give examples of how a person's demographic characteristics influence what he or she buys.

Social Studies Connection: Have students choose one of these positions: "We should ban junk food in schools" or "We should not ban junk food in schools." Group students in teams of three or four based on their position. Tell students that they will research their position and create a multimedia presentation that argues their position. Hold a class discussion about fact-based versus opinion-based arguments. Create a T-chart with one column labeled *fact* and the other labeled *opinion*. Ask students the following questions, recording their responses on the chart:

- What is a fact?

- What is an opinion?
- What is the difference between an argument based on fact and an argument based on opinion?

The class should conclude that a *fact* is a detail that can be proven true or false while an *opinion* is an expression of feelings. It is important for students to understand that opinions can be based on facts, but that the opinion itself cannot be proven to be true or false. You may wish to provide some examples of facts versus opinions such as "George Washington was the first president of the United States" (fact) and "Abraham Lincoln was a great president" (opinion).

Provide students with the Should We Ban Junk Food in Schools? Multimedia Presentation Rubric (Rubric 4 on p. 90), and go over the expectations for the presentation with the class. Have students spend day 4 of the lesson working on their presentations. On day 5, students should finalize and then give their multimedia presentations on whether junk food should be banned in schools.

Elaboration/Application of Knowledge

Science Connection: Students should gather their materials and begin to construct their containers for the Save the Chips Competition. Student teams need to conduct several tests to ensure that their chip containers will withstand the various conditions described.

STEM Resource Notebook Prompt

Have students respond to the following STEM Research Notebook prompt near the conclusion of the day.

- *What solutions did you pursue today?*
- *Which designs were successful?*
- *Which designs were unsuccessful? Why?*
- *How will you use what you learned today to help improve your design tomorrow?*

During the next class session, have students share their STEM Research Notebook responses with their peers. Ask them how they can improve their design to have the smallest number of broken chips. Students should then continue to revise and retest their designs for the Save the Chips Competition. At the end of class, inform students that in the next class, they will hold their final Save the Chips Competition.

The next day, have students hold the competition by conducting throw tests, crush tests, and drop tests. The team with the smallest number of broken chips wins the competition.

STEM Research Notebook Prompt

Have students answer the following questions in their STEM Research Notebooks:

- *What did you learn about your design, and what made it effective or ineffective at saving the chips?*
- *How did you use the engineering design process to solve a real-world problem?*

Reintroduce the Product Design Challenge (see Product Design Challenge Student Handout) to students. Ask them to write in their STEM Research Notebooks how they will use the engineering design process to meet this challenge. How can they use what they learned in the Save the Chips Competition to help them create better packaging for an existing product in the module challenge? Have a few students share their STEM Research Notebook entries and discuss as a class.

Mathematics Class: Remind students that in the last math class, they were asked to find the number of broken chips in a bag. As a class, discuss whether the number of broken chips they found was acceptable to them as consumers. Tell students that, working as groups, they are going to develop surveys that will allow them to explore the school's response to the number of broken chips in each bag. Have students work in teams of three to four to develop a research question and devise a plan that allows them to answer the question. Conference with student groups as they create their questions and surveys. Guide them as necessary using the following questions:

- Is your question researchable?
- What data will you need to collect to answer the question?
- Whom will you ask? How many people would you need to ask?
- What instrument will you need to develop to answer your question?
- How will you display your results and analyze your data?
- How else will you present your findings?

Have each team create a miniposter of their research plan to share with their peers. Allow students to give feedback on their peers' plans. Inform students that they should collect data before the next class as homework.

During the next class session, have students analyze their data and create a representation that displays their findings. Then ask students to write a letter to the company in which they share those findings. Provide them with the Letter to the Company Rubric (Rubric 3, pp. 88–89). An additional option for this activity is to have students exchange letters and create response letters, acting as customer service representatives of the company.

The following day, review the three-act lesson again. Review the content of the nutritional information printed on the chip packages, pointing out calories, serving size, and servings per bag. Have students calculate the number of calories, grams of fat, and milligrams of sodium they would consume if they ate an entire small bag, medium bag, or large bag of tortilla chips. Student teams should also begin to work together to choose their products for the module challenge.

ELA Class: Divide the class into five discussion groups, and give each a card with one of the demographic categories, a sheet of chart paper on which to record their brainstormed ideas, and a marker. Have each group brainstorm ways that its category might affect purchasing choices. Each group must come up with at least three plausible connections between demographics and purchasing. Have students share and discuss their ideas. Reiterate to the class that who we are and how we live affects what we buy and that marketers trying to persuade people to buy their products try to send messages that appeal to certain kinds of buyers.

Working in pairs or small groups, students should choose one product package from those provided and try to create a profile of the targeted consumer using demographic characteristics and other buying factors. Guide students by asking these questions:

- What is the age of your targeted consumer?
- What is the gender of your targeted consumer?
- In what kind of family does your targeted consumer live? What role does your consumer play in the family?
- What is the cultural identity or identities of your consumer? Consider nationality, ethnic group, and language.
- Does this person make a lot of money? Does this person have control over the spending for the household?
- What does this person like to do?
- What else do you need to consider?

For example, a group analyzing a box of cereal labeled as organic, whole grain, and lightly sweetened might construct the following consumer profile: The cereal seems to be marketed to adults, because it doesn't use flashy colors or cartoon images, but it would appeal to adults who are hoping to get their kids to eat healthy food, especially since the cereal is sweetened. "Lightly sweetened," "whole grain," and "organic" all would appeal to someone who is health conscious, so this person might have an active lifestyle. The box notes that "sustainable practices" are used in growing the ingredients, so that would appeal to people who are concerned about the environment, possibly someone who enjoys outdoor activities like hiking. The targeted consumer probably makes a lot of money, too, because organic foods are often priced higher than other foods. People who buy them usually have enough money to have a lot of choices about what they buy. There isn't a clear ethnic or cultural link to this food; people of many different cultures within the United States might buy it and enjoy it. Dry cereal is a quick option for breakfast, so this product might appeal to busy, working people. Healthy cereals are often located together in one place on the cereal aisle, which means a consumer who buys this is probably searching for it on purpose, so it doesn't need to have flashy colors like regular cereal boxes.

Social Studies Connection: Discuss the following with the class: Blogs are an effective way to get information out quickly to the public. However, some blogs are full of misinformation and opinion and are not always based on fact. Point out to students that it is important to always fact-check things they read online.

Have students read the article by Adam on the blog *Why This Food Is Garbage!* at *http://whythisfoodisgarbage.blogspot.com/2011/01/doritos.html.* Students should explore the information presented in the blog by conducting additional research about whether the information is true or false and reflects fact or opinion. Also have them investigate the question, What are the impacts of monosodium glutamate (MSG) and trans fats on our bodies?

Have students write a response in their STEM Research Notebooks that refutes or supports Adam's position on Doritos in his blog. In their response, students need to provide information that they learned from their research on junk food.

As a class, explore the economic impact of the Doritos Locos Tacos on Taco Bell's bottom line, and then discuss the following questions about junk food:

- Do you think people in the United States are obsessed with junk food?
- What about people in other parts of the world?
- What is the economic impact of the obsession with junk food in the United States?
- What are other impacts of America's obsession with junk food?

STEM Resource Notebook Prompt

Have students respond to the following questions in their STEM Research Notebooks, and then discuss as a class:

- *Describe the impacts of food deserts and food swamps on middle school students.*
- *Do you think the school should be responsible for providing students with fresh fruits and vegetables?*

- *Do you think there is a time and place when junk food should be allowed?*
- *How does marketing and packaging play into our obsession with junk food?*

Evaluation/Assessment

Students may be assessed on the following performance tasks and other measures listed.

Performance Tasks

- Rubric 2: Save the Chips Competition Design Rubric
- Rubric 3: Letter to the Company Rubric
- Rubric 4: Should We Ban Junk Food in Schools? Multimedia Presentation Rubric
- Rubric 5: Response to Blog Rubric

Other Measures

- STEM Research Notebook entries
- Participation in class discussions

INTERNET RESOURCES

Background on the EDP

- *www.sciencebuddies.org/engineering-design-process/engineering-design-process-steps.shtml*
- *www.eie.org/overview/engineering-design-process*
- *www.teachengineering.org/engrdesignprocess.php*
- *http://curriculum.vexrobotics.com/curriculum/intro-to-engineering/what-is-the-engineering-design-process*

Ideation sessions

- *www.interaction-design.org/literature/article/how-to-select-the-best-idea-by-the-end-of-an-ideation-session*

Dan Meyer's blog on Three-Act Tasks

- *http://blog.mrmeyer.com/2013/teaching-with-three-act-tasks-act-one*

Career information

- *www.bls.gov/ooh/architecture-and-engineering/chemical-engineers.htm*
- *www.bls.gov/ooh/architecture-and-engineering/mechanical-engineers.htm*

- *www.bls.gov/ooh/life-physical-and-social-science/physicists-and-astronomers.htm*
- *www.bls.gov/ooh/math/mathematicians-and-statisticians.htm*

Media literacy

- *www.medialit.org/media-literacy-definition-and-more*
- *www.medialit.org/reading-room/five-key-questions-form-foundation-media-inquiry*
- *www.medialit.org/reading-room/media-literacy-middle-school*

Using the *New York Times* to teach text type

- *http://learning.blogs.nytimes.com/2011/12/12/compare-contrast-cause-effect-problem-solution-common-text-types-in-the-times*

Problem solving with flow charts

- *https://sciencing.com/solve-math-problems-using-flowchart-7840920.html*

Reading and writing nonfiction

- *www.learner.org/workshops/teachreading35/classrooms/cv8.html*
- *www.learner.org/courses/readwrite/video-detail/organizing-ideas-multiple-sources.html*
- *www.learner.org/courses/readwrite/video-detail/teaching-content-through-literacy.html*
- *www.learner.org/workshops/writing35/session8/index.html*
- *www.nwp.org/cs/public/print/resource_topic/teaching_writing*

Information on food deserts and food swamps

- *http://mic.com/articles/7176/obesity-food-deserts-have-given-way-to-food-swamps*
- *http://voices.washingtonpost.com/all-we-can-eat/food-politics/food-deserts-vs-swamps-the-usd.html*
- *http://brownisthenewpink.com/2014/05/27/food-deserts-and-swamps-social-justice-issue*
- *https://www.theatlantic.com/health/archive/2017/12/food-swamps/549275*

"Deep Inside Taco Bell's Doritos Locos Taco"

- *www.fastcompany.com/3008346/deep-inside-taco-bells-doritos-locos-taco*

New York Times "Fixes" blog

- *http://opinionator.blogs.nytimes.com/category/fixes*

"Toys & Games TV Commercials"

- *www.ispot.tv/browse/w.dQ/life-and-entertainment/toys-and-games*

Article on *Why This Food Is Garbage!* blog

- *http://whythisfoodisgarbage.blogspot.com/2011/01/doritos.html*

"How Tortilla Chips Are Made" video

- *www.youtube.com/watch?v=QzIdZGOR9vo*

"Tortilla Chips Act One" video

- *www.youtube.com/watch?v=TbO79YIBu00*

"Nonstick Chewing Gum"

- *http://sciencenetlinks.com/science-news/science-updates/nonstick-chewing-gum*

"This Sobering Map Shows You All of America's Food Deserts"

- *https://grist.org/food/this-sobering-map-shows-you-all-of-americas-food-deserts*

ENGINEERING DESIGN PROCESS

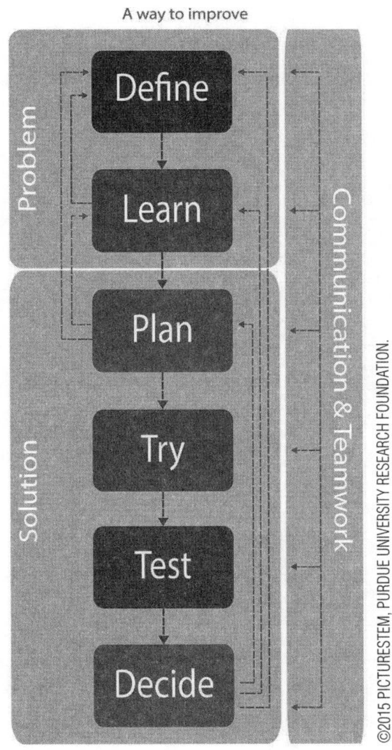

Name: ______________________________

STUDENT HANDOUT

TOO MANY BROKEN CHIPS EXPLORATION SHEET

How many chips are in your bag?

What fraction of the chips were broken? Write this as a percentage. Explain how you found your answer using pictures, words, and symbols.

As a group, calculate the average number of broken chips.

What is the class average number of broken chips? How did you figure this out? Do you think this an acceptable number? What do you think other students at your school will say is acceptable? How could we find an answer to that question?

Name: ______________________________

STUDENT HANDOUT

PROBLEM-SOLUTION GRAPHIC ORGANIZER

Problem

Potential Solutions to the Problem

What solutions did you attempt?	What were the results of those attempts?

Best Solution and Why

Name: ______________________________

STUDENT HANDOUT, PAGE 1

THREE-ACT LESSON IMAGES

PHOTOS BY ADRIENNE REDMOND-SANOGO

ACT-TWO IMAGES

Outside of Large Package

Outside of Small Package

Inside of Large Package

Inside of Small Package

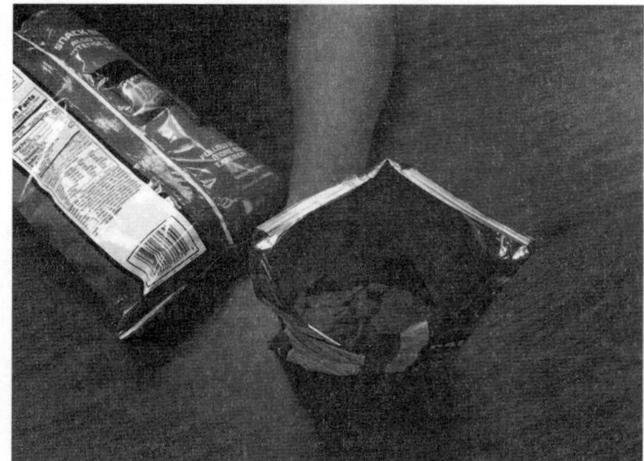

Name: ____

STUDENT HANDOUT, PAGE 2

THREE-ACT LESSON IMAGES

PHOTOS BY ADRIENNE REDMOND-SANOGO

ACT-TWO IMAGES

Nutritional Information on a Chip Package

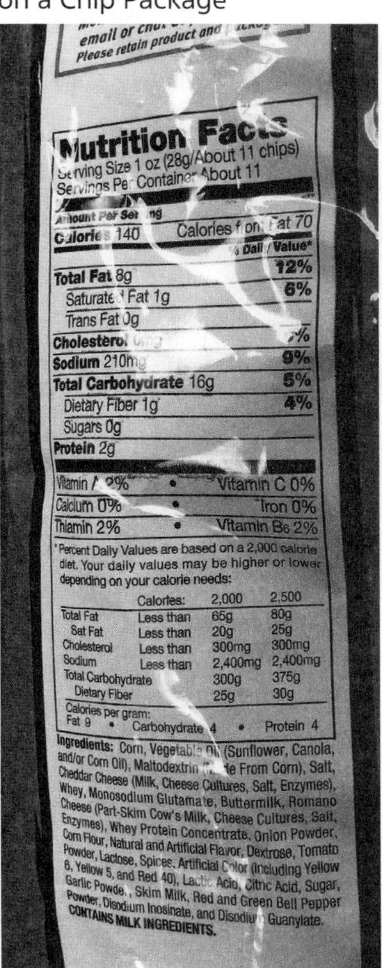

Large and Small Packages Poured Out

Name: ______________________________

STUDENT HANDOUT, PAGE 3

THREE-ACT LESSON IMAGES

PHOTOS BY ADRIENNE REDMOND-SANOGO

ACT-THREE IMAGES

Contents of Small Bag

Contents of Large Bag

Rubric 1: Product Design Challenge Rubric										
Name ____________										
Product Selected										
Product selected clearly meets the criteria, and the evidence is clearly articulated in the presentation.										
YES			MOSTLY			SOMEWHAT				NOT AT ALL
10	9	8	7	6	5	4	3	2	1	0
Packaging Design										
Packaging design clearly meets the criteria, and the evidence is clearly articulated in the presentation.										
YES			MOSTLY			SOMEWHAT				NOT AT ALL
10	9	8	7	6	5	4	3	2	1	0
Marketing Campaign										
Marketing campaign clearly meets the criteria, and the evidence is clearly articulated in the presentation.										
YES			MOSTLY			SOMEWHAT				NOT AT ALL
10	9	8	7	6	5	4	3	2	1	0
Product testing clearly matches the criteria, and the evidence is clearly articulated in the presentation.										
YES			MOSTLY			SOMEWHAT				NOT AT ALL
10	9	8	7	6	5	4	3	2	1	0
Presenters										
Presenters appear to be well prepared, speak clearly and coherently, and make eye contact.										
YES			MOSTLY			SOMEWHAT				NOT AT ALL
10	9	8	7	6	5	4	3	2	1	0
Presentation										
Presentation is energetic, creative, and engaging.										
YES			MOSTLY			SOMEWHAT				NOT AT ALL
10	9	8	7	6	5	4	3	2	1	0
COMMENTS:										

Rubric 2: Save the Chips Competition Design Rubric

Name ____________________

Criteria	1 Below Mastery	2 Approaching Mastery	3 At Mastery	4 Advanced
PROBLEM CLARIFICATION	Students misunderstood the problem or didn't clarify the problem in the presentation.	Students' understanding of the problem is shallow or not clarified well in the presentation.	Students' understanding of the problem is evident in that problem is discussed and reworded. However, students may not use technical language when describing the problem.	Students' understanding of the problem is evident in that the problem is discussed and reworded systematically. Students use technical language accurately when describing the problem. Students were able to analyze how the parts of the whole interacted to produce the overall complex solution.
DESIGN CONCEPT	The design does not show use of the EDP. The design is not effective and does not meet all the requirements for the competition.	The design shows little use of the EDP. The design is not effective, but students attempted to meet the requirements for the competition.	The design uses the EDP. The design is effective, and students met the requirements for competition.	The design is sophisticated and shows that students refined their initial design to make it more appealing and effective. Students met the requirements to create an effective product.
COMMUNICATION	Students were not able to effectively communicate their design process and describe their design's successes and failures during their presentation.	Students explained their ideas but may not have been effective at sharing those ideas. Students did not attempt to listen and respond to questions during their presentation.	Students were able to explain their thoughts and ideas using oral, written, or nonverbal communication skills but not all three. Students were able to listen to and respond effectively to questions posed during their presentation.	• Students were able to clearly and effectively articulate their thoughts and ideas using oral, written, and nonverbal communication skills. • Students were able to listen to and respond effectively to questions posed during their presentation.

Continued

Rubric 2: Save the Chips Competition Design Rubric (*continued*)

Name ______________________

Criteria	1 Below Mastery	2 Approaching Mastery	3 At Mastery	4 Advanced
GROUP COLLABORATION AND COOPERATION	Students did not demonstrate the ability to work effectively and respectfully with their team. They did not work together well and failed to make compromises for the good of the project. They did not respect the contributions of the team.	Students struggled to work effectively and respectfully with their team. However, they did attempt to make compromises and listen to their teammates.	Students were able to work effectively as a team. They made compromises at times but may have had difficulty respecting the contributions of their teammates.	• Students demonstrated the ability to work effectively and respectfully with their team. • They exercised flexibility and willingness to help in making necessary compromises to accomplish a common goal. • They assumed shared responsibility for collaborative work and valued contributions made by each team member.
COMMENTS:				

Rubric 3: Letter to the Company Rubric

Name: ______________________

Criteria	1 Below Mastery	2 Approaching Mastery	3 At Mastery	4 Advanced
IDENTIFICATION OF THE ISSUE	Student's opinion on the issue is not clearly stated, shows a lack of understanding, and is not related to the need to create better packaging to prevent broken chips.	Student's opinion is not clearly stated and has little reference to the issue of broken chips.	Student's personal opinion is clearly stated and related to the issue of broken chips.	Student's personal opinion about broken chips is strongly and clearly stated and identifies the issue.
ARGUMENT AND SUPPORT	The student's argument is weak and makes fewer than three points.	The student provides two points to support the need for better packaging, but the arguments are weak.	Student provides three or more points to support the need for better chip packaging, but the arguments are somewhat weak in spots. The letter is not strongly persuasive.	Student provides three or more excellent points with strong evidence to support the need for better packaging of chips. It is evident that the writer put thought and effort into the letter.
CONCLUDING STATEMENT	The conclusion does not refer to the student's opinion about the number of broken chips.	The conclusion is a weak summary of the student's opinion about the number of broken chips.	The conclusion summarizes the student's opinion about the number of broken chips.	The conclusion provides a strong summary of the student's opinion about the number of broken chips and the need for improved packaging.
ORGANIZATION AND CLARITY	The organization of the letter is confusing and lacks proper sentence and paragraph structure.	The letter contains inconsistent use of paragraph structure.	The organization of the letter is clear and uses proper paragraph structure.	The organization of the letter is logical and well thought out. The letter contains well-written and varied sentences and proper paragraph structure.

Continued

Rubric 3: Letter to the Company Rubric (*continued*)

Name: ____________________

Criteria	1 Below Mastery	2 Approaching Mastery	3 At Mastery	4 Advanced
WORD CHOICE AND TONE	The language of the letter is unclear and lacks description. The letter may have offensive or inappropriate words or tone.	The letter contains some words that are clear and descriptive but lacks a consistent persuasive tone.	The letter uses an adequate choice of words that are clear and descriptive. Parts of the letter demonstrate a persuasive tone.	The letter contains a choice of words that are clear, descriptive, and accurate. The letter maintains a persuasive tone throughout.
MECHANICS AND GRAMMAR	The letter contains many punctuation, spelling, and grammatical errors that interfere with the meaning and purpose of the letter.	The letter contains several punctuation, spelling, and grammatical errors that interfere with the meaning and purpose of the letter.	The letter contains some errors in punctuation, spelling, and grammar that do not interfere with the meaning and purpose of the letter.	The letter contains few or no punctuation, spelling, and grammatical errors.

COMMENTS:

Rubric 4: Should We Ban Junk Food in Schools? Multimedia Presentation Rubric

Name: ______________________

Criteria	1 Below Mastery	2 Approaching Mastery	3 At Mastery	4 Advanced
EXPLANATION	Students use too few, inappropriate, or irrelevant descriptions, facts, details, or examples to support ideas.	Students use some descriptions, facts, details, and examples that support ideas, but there may not be enough or some may be irrelevant.	Students use relevant, descriptions, facts, details, and examples to support claims, findings, and arguments.	Students use relevant, well-chosen descriptions, facts, details, and examples to create a compelling and convincing argument.
ORGANIZATION	Presentation lacks important required parts, such as an introduction, main idea, or conclusion. It presents ideas in an order that does not make sense. Students use time poorly; the whole presentation, or a part of it, is too short or too long.	Presentation includes almost everything required. It moves from one idea to the next, but the main idea may not be clear, or some ideas may be in the wrong order. It has an introduction and a conclusion, but they are not effective. Time is generally used well, but students may spend too much or too little time on a topic, idea, or audio-visual aid.	Presentation includes everything required. It includes an effective introduction, a clearly stated main idea, and a conclusion. Students organize time well; no part is rushed, too short, or too long.	Presentation includes everything required with an outstanding organizational flow. Presentation includes an effective introduction, a clearly stated main idea, and a conclusion, and students move from one idea to the next in a logical order, emphasizing main points in a focused, coherent manner. Students organize time well; no part is rushed, too short, or too long.
PRESENTATION AIDS	Students do not use audio-visual aids or media, or they attempt to use one or a few audio-visual aids or media, but these distract from or do not add to the presentation.	Students use audio-visual aids or media, but these sometimes distract from or do not add to the presentation.	Students use well-produced audio-visual aids or media to add interest.	Students use well-produced audio-visual aids or media to clarify information, emphasize important points, strengthen arguments, and add interest.

Continued

Rubric 4: Should We Ban Junk Food in Schools? Multimedia Presentation Rubric (*continued*)

Name: ______________________

Criteria	1 Below Mastery	2 Approaching Mastery	3 At Mastery	4 Advanced
RESPONSE TO QUESTIONS	Students do not address audience questions They may go off topic or misunderstand without seeking clarification.	Students answer some audience questions, but not always clearly or completely.	Students answer audience questions clearly and completely but may not seek clarification, admit they don't know, or explain how the answer might be found when unable to answer a question.	Students answer audience questions clearly and completely. When necessary, they seek clarification, admit they don't know, or explain how the answer might be found when unable to answer a question.
PARTICIPATION	Not all team members participate; only one or two speak.	All team members participate, but not equally.	All team members participate for about the same length of time.	All team members participate for about the same length of time. All team members are able to answer questions about the topic as a whole, not just their part of it.
COMMENTS:				

Rubric 5: Response to Blog Rubric

Name: ____________________

Criteria	1 Below Mastery	2 Approaching Mastery	3 At Mastery	4 Advanced
ORGANIZATION	Student's response is unorganized and scattered or includes a phrase or one-word comment (e.g., "Yes! I agree with the author.").	Student's response is somewhat disconnected with little focus.	Student's response is somewhat concise and focused.	Student's response is concise and focused.
SUPPORT FOR ARGUMENT	Student provides no support for the argument.	Student states only opinions without providing appropriate support, or support provided is unrelated.	Student supports or challenges the blog post author's position by citing facts in support of his or her position.	Student effectively and respectfully supports or challenges the blog post author's position by citing facts in support of his or her position.
UNDERSTANDING OF TOPIC	Student does not demonstrate understanding of the blog topic.	Student demonstrates a weak understanding of the blog topic.	Student demonstrates an understanding of the blog topic.	Student demonstrates detailed understanding of the blog topic.
SPELLING AND GRAMMAR	Response contains multiple spelling and grammar errors that distract from its readability.	Response contains a few spelling and grammar errors, but these do not distract from its readability.	Response contains no more than three errors in spelling and grammar.	Response contains no errors in spelling and grammar.
COMMENTS:				

STUDENT HANDOUT, PAGE 1

PRODUCT DESIGN CHALLENGE GUIDELINES

Welcome to the Product Design Challenge! Working with a team of your peers, you will redesign a familiar product and create new packaging and a marketing campaign, resulting in a hot new product for sale by your company.

GUIDELINES

PRODUCT: Choose a familiar product with potential to be redesigned for marketing to a new user, thus expanding the market base and company revenue. Product requirements:

1. The product should be useful or highly appealing to a particular consumer group, or both.
2. It should be something that is fragile enough to need protection during shipping.

PACKAGING: Using the engineering design process (EDP), your team will design packaging for both shipping and display of your product. Packaging requirements:

1. Your package design should match consumer expectations, yet be different in some way from similar packaging.
2. Your packaging should use materials already produced by your company, so that the cost is minimized.
3. Your packaging should have a design that is unique and bold to attract buyers in your target market.
4. Your packaging should be sustainable and environmentally friendly.
5. Your packaging should be strong and sturdy enough to ship and display on store shelves without damaging the product.

MARKETING: Design a highly effective marketing strategy, including a product logo and marketing campaign. Marketing requirements:

1. Your marketing should feature a logo that is highly appealing, unique, bold, and memorable in order to attract and retain buyers
2. Your marketing campaign should tell a story, have a focus, and lead consumers on a journey to where you want them to go.

STUDENT HANDOUT, PAGE 2

PRODUCT DESIGN CHALLENGE GUIDELINES

3. Your marketing campaign should include messaging that effectively works to attract consumers across a variety of market media, such as when displayed in stores, seen online, or featured in a magazine advertisement.

PRODUCT QUALITY TESTING: You need to test your product for reliability and appeal, and then redesign as needed. Product quality testing requirements:

1. Your packaging must pass a crush test designed to demonstrate that the product is adequately protected.
2. You will present your product to a group of potential consumers (another team in your class), and their feedback must demonstrate that your product has appeal to them.

SELLING YOUR INNOVATION: You will present your product, packaging design, and marketing campaign to a panel of "company executives" (school administrators and community members your teacher invites to act as company executives). The presentations will be judged on the following criteria:

1. The product selected clearly meets the requirements, and the evidence is clearly articulated in the presentation.
2. The packaging design clearly meets the requirements, and the evidence is clearly articulated in the presentation.
3. The marketing campaign and logo clearly meet the requirements, and the evidence is clearly articulated in the presentation.
4. The product testing shows that the product quality clearly meets the requirements, and the evidence is clearly articulated in the presentation.
5. The presenters appear to be well prepared, speaking clearly and coherently and making eye contact, and the presentation is energetic, creative, and engaging.

Lesson Plan 2: The Packaging

In this lesson, students continue to prepare for their Product Design Challenge by building background knowledge required to complete the project, including exploring three-dimensional shapes, calculating surface area and volume, and learning about packaging materials and sustainability. Students repurpose their product's packaging and prepare for the marketing pitch they will develop in Lesson 3. Finally, students explore the social justice issues associated with product manufacturing and packaging.

ESSENTIAL QUESTIONS

- How are engineering design skills and communication skills employed in the design and marketing of consumer products?
- Where do packages and products originate, and how can manufacturers make sure that their products and packaging are sustainable?
- How are surface area and volume used to create packaging?
- What are the social justice issues associated with manufacturing and packaging products?
- Who creates media messages? For what purposes?
- What techniques do creators of media messages use to attract our attention?
- How might different people understand and respond to media messages differently?
- What perspectives are represented in various media messages? From what perspective do you view them, and how does this affect your understanding and response?
- How can readers and writers use nonfiction text structures and features to comprehend and compose informational texts?
- How do writers compose effective argumentative and persuasive messages?

ESTABLISHED GOALS AND OBJECTIVES

At the conclusion of this lesson, students will be able to do the following:

- Understand where products and packaging originate and end up (life cycle of a product)
- Discuss the sustainability issues associated with packaging and manufacturing of products

- Use surface area and volume to design packaging
- Calculate surface area and volume of three-dimensional figures, and develop a general formula
- Comprehend and produce informational texts for authentic purposes
- Summarize and synthesize information from multiple texts
- Critically analyze different kinds of texts (print and multimedia) for structures and techniques that message creators use to communicate and influence
- Understand that individuals interpret texts—both print and media—differently, depending on their experiences and perspectives
- Prepare and perform commercial scripts
- Use a variety of media (e.g., print, art, video) to communicate complex information
- Use oral and written language effectively to collaborate and problem solve in a work community context

TIME REQUIRED

- 6 days (approximately 45 minutes each day; see Tables 3.7–3.8, pp. 40–41)

MATERIALS

Required Materials for Lesson 2

- STEM Research Notebooks
- Computer and projector for students to watch videos and share ideas
- Computers, tablets, or laptops with internet access for student research and presentations
- Handouts (attached at the end of this lesson)
- Chart paper

Additional Materials for Mathematics Class

- Note cards
- Packages in a variety of three-dimensional shapes (3–5 types of packages per group of 3–4 students; see p. 113 for more details)

- Three-dimensional objects and manipulatives, preferably with nets (two-dimensional shapes that can be folded to create three-dimensional shapes) (6–8 per group of 3–4 students)
- Unit cubes in a variety of sizes (3–5 per group of 3–4 students)
- Three-dimensional geometric shapes, preferably with nets (1 set per group)
- Rulers (1 per group of 3-4 students)
- Digital camera (optional)

Additional Materials for Science Connection

- Poster board
- Markers

Additional Materials for ELA Class

- Sticky notes

CONTENT STANDARDS AND KEY VOCABULARY

Table 4.4 lists the content standards from the *NGSS, CCSS,* and the Framework for 21st Century Learning that this lesson addresses, and Table 4.5 (p. 101) presents the key vocabulary. Vocabulary terms are provided for both teacher and student use. Teachers may choose to introduce some or all of the terms to students.

Table 4.4. Content Standards Addressed in STEM Road Map Module Lesson 2

NEXT GENERATION SCIENCE STANDARDS
PERFORMANCE EXPECTATIONS
• MS-ETS1-1. Define the criteria and constraints of a design problem with sufficient precision to ensure a successful solution, taking into account relevant scientific principles and potential impacts on people and the natural environment that may limit possible solutions.
• MS-ETS1-2. Evaluate competing design solutions using a systematic process to determine how well they meet the criteria and constraints of the problem.
• MS-ETS1-3. Analyze data from tests to determine similarities and differences among several design solutions to identify the best characteristics of each that can be combined into a new solution to better meet the criteria for success.
• MS-PS1-3. Gather and make sense of information to describe that synthetic materials come from natural resources and impact society.
• MS-ESS3-3. Apply scientific principles to design a method for monitoring and minimizing a human impact on the environment.

Continued

Table 4.4. (*continued*)

DISCIPLINARY CORE IDEAS

ETS1.A. Defining and Delimiting Engineering Problems

- The more precisely a design task's criteria and constraints can be defined, the more likely it is that the designed solution will be successful. Specification of constraints includes consideration of scientific principles and other relevant knowledge that are likely to limit possible solutions.

ETS1.B. Developing Possible Solutions

- There are systematic processes for evaluating solutions with respect to how well they meet the criteria and constraints of a problem.
- Sometimes parts of different solutions can be combined to create a solution that is better than any of its predecessors.

ETS1.C. Optimizing the Design Solution

- Although one design may not perform the best across all tests, identifying the characteristics of the design that performed the best in each test can provide useful information for the redesign process—that is, some of those characteristics may be incorporated into the new design.

ESS3.C. Human Impacts on Earth Systems

- Human activities have significantly altered the biosphere, sometimes damaging or destroying natural habitats and causing the extinction of other species. But changes to Earth's environments can have different impacts (negative and positive) for different living things.

CROSSCUTTING CONCEPTS

Influence of Science, Engineering and Technology on Society and the Natural World

- All human activity draws on natural resources and has both short- and long-term consequences, positive as well as negative, for the health of people and the natural environment.
- The uses of technologies and any limitations on their use are driven by individual or societal needs, desires, and values; by the findings of scientific research; and by differences in such factors as climate, natural resources, and economic conditions. Thus technology use varies from region to region and over time.

Cause and Effect

- Relationships can be classified as causal or correlational, and correlation does not necessarily imply causation.

Structure and Function

- Structures can be designed to serve particular functions by taking into account properties of different materials, and how materials can be shaped and used.

Interdependence of Science, Engineering, and Technology

- Engineering advances have led to important discoveries in virtually every field of science, and scientific discoveries have led to the development of entire industries and engineered systems.

Continued

Table 4.4. (*continued*)

SCIENCE AND ENGINEERING PRACTICES

Obtaining, Evaluating, and Communicating Information

- Obtaining, evaluating, and communicating information in 6–8 builds on K–5 and progresses to evaluating the merit and validity of ideas and methods.
- Gather, read, and synthesize information from multiple appropriate sources and assess the credibility, accuracy, and possible bias of each publication and methods used, and describe how they are supported or not supported by evidence.

COMMON CORE STATE STANDARDS FOR MATHEMATICS

MATHEMATICAL PRACTICES

- MP1. Make sense of problems and persevere in solving them.
- MP3. Construct viable arguments and critique the reasoning of others.
- MP4. Model with mathematics.
- MP5. Use appropriate tools strategically.
- MP6. Attend to precision.

MATHEMATICAL CONTENT

- 6.RP.A.3.C. Find a percent of a quantity as a rate per 100 (e.g., 30% of a quantity means 30/100 times the quantity); solve problems involving finding the whole, given a part and the percent.
- 6.SP.A.1. Recognize a statistical question as one that anticipates variability in the data related to the question and accounts for it in the answers.
- 6.SP.A.3. Recognize that a measure of center for a numerical data set summarizes all of its values with a single number, while a measure of variation describes how its values vary with a single number.
- 6.SP.B.5.B. Describing the nature of the attribute under investigation, including how it was measured and its units of measurement.
- 6.SP.B.5.C. Giving quantitative measures of center (median and/or mean) and variability (interquartile range and/or mean absolute deviation), as well as describing any overall pattern and any striking deviations from the overall pattern with reference to the context in which the data were gathered.

COMMON CORE STATE STANDARDS FOR ENGLISH LANGUAGE ARTS

READING STANDARDS

- RH.6–8.4. Determine the meaning of words and phrases as they are used in a text, including vocabulary specific to domains related to history/social studies.
- RI.6.2. Determine a central idea of a text and how it is conveyed through particular details; provide a summary of the text distinct from personal opinions or judgments.

Continued

Table 4.4. (*continued*)

READING STANDARDS (*continued*)

- RI.6.6. Determine an author's point of view or purpose in a text and explain how it is conveyed in the text.

WRITING STANDARDS

- W.6.4. Produce clear and coherent writing in which the development, organization, and style are appropriate to task, purpose, and audience.
- W.6.6. Use technology, including the Internet, to produce and publish writing as well as to interact and collaborate with others; demonstrate sufficient command of keyboarding skills to type a minimum of three pages in a single sitting.
- WHST.6–8.4. Produce clear and coherent writing in which the development, organization, and style are appropriate to task, purpose, and audience.
- WHST.6–8.6. Use technology, including the Internet, to produce and publish writing and present the relationships between information and ideas clearly and efficiently.
- WHST.6–8.8. Gather relevant information from multiple print and digital sources, using search terms effectively; assess the credibility and accuracy of each source; and quote or paraphrase the data and conclusions of others while avoiding plagiarism and following a standard format for citation.
- WHST.6.10. Write routinely over extended time frames (time for reflection and revision) and shorter time frames (a single sitting or a day or two) for a range of discipline-specific tasks, purposes, and audiences.

FRAMEWORK FOR 21ST CENTURY LEARNING

21st Century Themes

- Environmental Literacy
- Health Literacy
- Global Awareness
- Financial, Economic, Business, and Entrepreneurial Literacy
- Civic Literacy

Learning and Innovation Skills

- Creativity and Innovation
- Critical Thinking and Problem Solving
- Communication and Collaboration

Continued

Table 4.4. (*continued*)

FRAMEWORK FOR 21ST CENTURY LEARNING (*continued*)
Information, Media, and Technology Skills • Information Literacy • Media Literacy • ICT Literacy ***Life and Career Skills*** • Flexibility and Adaptability • Initiative and Self-Direction • Social and Cross Cultural Skills • Accountability, Leadership, and Responsibility

Table 4.5. Key Vocabulary in Lesson 2

Key Vocabulary	Definition
bioplastic	a biodegradable plastic substance made from biological materials
chief executive officer (CEO)	the head of a company, responsible for making important financial and managerial decisions
efficient	achieving maximum results with little waste of time or energy
environment	the natural world, including the land, water, and air, in which humans, animals, and plants live
environmental stewardship	responsibility for environmental quality, protecting the environment through conservation, restoration, and sustainable practices
landfill	a waste disposal site where waste materials are buried and covered with soil
line of symmetry	the imaginary line where you could fold an image or cut it into halves that would be mirror images of each other
nested packaging	packages within packages
net	a pattern that can be cut and folded to make a model of a solid shape
packaging	the container or wrapper used to protect or enclose a product

Continued

Table 4.5. (*continued*)

Key Vocabulary	Definition
packaging life cycle	the series of changes that packaging goes through, from its development from raw materials to decomposition in a landfill or use for a recycled purpose
plane of symmetry	the imaginary plane where you could cut a three-dimensional shape into halves that would be mirror images of each other
properties	characteristics or qualities; in shapes, this includes the number of sides, relative length of sides, whether opposite sides are parallel, and whether all angles are the same
recyclable	able to be recycled
recycle	convert waste products into reusable material
renewable	not depleted when used; capable of being replaced naturally
repurpose	change or adapt to use for a different purpose
social justice	equal rights and opportunities and just treatment for all humans in a society
sustainability	production of goods and services in a way that avoids depleting natural resources or harming the environment
sustainable	having a minimal impact on the environment
surface area	the total area of the surface of an object
three-dimensional	having height, width, and depth
volume	the amount of space occupied by a three-dimensional object, calculated by multiplying the object's length times width times height

TEACHER BACKGROUND INFORMATION

In this lesson, students learn about packaging, surface area, volume, and sustainability issues associated with packaging. The information and links in this section will help you engage students in this lesson.

Sustainability and Sustainable Packaging

LandLearn NSW (2013) defines sustainability as the "ability or capacity of something to be maintained or to sustain itself. It's about taking what we need to live now, without jeopardizing the potential for people in the future to meet their needs." The concept of sustainable packaging is a growing concern around the globe as the number of people on our planet rises. According to the Sustainable Packaging Coalition (2011), sustainable packaging has the following characteristics:

- *Must be safe and healthy for individuals and communities throughout its life cycle*
- *Must meet market criteria for performance and cost*
- *Is sourced, manufactured, transported, and recycled using renewable energy*
- *Optimizes the use of renewable or recycled source materials*
- *Is manufactured using clean production technologies and best practices*
- *Is made from healthy materials throughout the life cycle*
- *Is physically designed to optimize materials and energy*
- *Is effectively recovered and utilized in biological and/or industrial closed loop cycles*

Sustainable packaging provides opportunities for manufacturers, shipping companies, and individuals to preserve our environment. The following websites provide information about sustainability and sustainable packaging:

- *https://sustainablepackaging.org/wp-content/uploads/2017/09/Definition-of-Sustainable-Packaging.pdf*
- *www.packworld.com/design/sustainability*
- *www.theguardian.com/sustainable-business/2014/jul/18/good-product-bad-package-plastic-recycle-mistakes*

Outsourcing

When companies in the United States and other developed countries outsource manufacturing processes to developing countries, they often bring jobs and resources with them. These jobs and their infrastructure can be a great boost for the economy of the developing country and can mean big profits for the company. Although outsourcing has benefits for both the company and the host country, it can have negative implications for the host country as well. Manufacturing waste, low wages, poor living conditions, and poor working conditions are just a few of the issues that can plague countries that receive outsourced jobs. Responsible business practices are a must in today's global society. Here are a few websites to help you learn more about issues of manufacturing and outsourcing:

- *www.crf-usa.org/bill-of-rights-in-action/bria-17-2-c-globalization-and-worker-rights*
- *https://globalexchange.org/campaigns/legacy-campaigns/fair-trade/*
- *www.slideshare.net/anilsural/is-outsourcing-exploitation-or-chance-for-developing-countries*
- *www.theguardian.com/commentisfree/2013/apr/29/bangladesh-factory-tragedy-sweatshop-economics*
- *www.chinalaborwatch.org/report/52*

Teaching About Social Justice

The following resources provide important information about how to teach children about social justice:

- Reach and Teach has a collection of definitions of *social justice* and provides some resources for teachers at *www.reachandteach.com/content/index.php?topic=socialjustice.*
- Education World offers "10 Social Justice Activities to Try in Class," with activities to teach elementary and middle grade students about social justice, at *www.educationworld.com/a_lesson/social-justice-activities-students.shtml.*
- We Are Teachers offers a list of picture books relating to social justice in "24 Books That Teach About Social Justice" at *www.weareteachers.com/books-about-social-justice.*

COMMON MISCONCEPTIONS

Students will have various types of prior knowledge about the concepts introduced in this lesson. Table 4.6 outlines some common misconceptions students may have concerning these concepts. Because of the breadth of students' experiences, it is not possible to anticipate every misconception that students may bring as they approach this lesson. Incorrect or inaccurate prior understanding of concepts can influence student learning in the future, however, so it is important to be alert to misconceptions such as those presented in the table.

Table 4.6. Common Misconceptions About the Concepts in Lesson 2

Topic	Student Misconception	Explanation
Sustainability	Sustainability only has environmental impacts.	Sustainability actually has impacts on society and the economy, in addition to the environment.
Area and volume	Area and volume mean the same thing.	Surface area and volume are two different measures of geometric shapes. *Area* is the amount of space inside of a flat, or two-dimensional, object's boundaries. The surface area of a three-dimensional shape is the sum of the areas of each surface of the object. *Volume* is the amount of space inside a three-dimensional object.

PREPARATION FOR LESSON 2

Review the Teacher Background Information provided (p. 103), assemble the materials for the lesson, and preview the videos recommended in the Learning Components section below.

For ELA class, identify several sources of information about youth marketing. Either print copies of information or ensure that students will have access to the internet. Each student pair will review two different resources. Some examples of youth marketing resources include the following:

- "Youth Marketing" at *www.marketing-schools.org/types-of-marketing/youth-marketing.html*
- CBS News, "Resources: Marketing to Kids" at *www.cbsnews.com/news/resources-marketing-to-kids*
- 4Imprint, "Marketing to Today's Youth" at *https://info.4imprint.com/wp-content/uploads/1P-13-0711-Succeeding.pdf*
- "Sugary Drink Facts: Advertising to Children and Teens Score 2014" at *http://sugarydrinkfacts.org/resources/SugaryDrinkFACTS_ReportSummary.pdf*
- Campaign for a Commercial-Free Childhood, "Marketing to Children Overview" at *www.commercialfreechildhood.org/resource/marketing-children-overview*

In the science connection, student teams will research major shipping organizations' sustainability initiatives. Each team will research sustainability initiatives for one of the

following: United States Postal Service (USPS), FedEx, and UPS. Either print resources on these organizations' sustainability initiatives or ensure that students have access to the initiative websites. Resources can be found by conducting an internet search using search terms such as "[organization name] environment," "[organization name] sustainability." For example:

- USPS sustainability at *https://about.usps.com/what-we-are-doing/green/welcome.htm*
- FedEx environmental sustainability at *www.fedex.com/cv/about/sustainability/index.html*
- UPS "The Power of More" at *https://sustainability.ups.com/committed-to-more*
- For social studies, prepare a lecture or discussion on the geography of marketing (see p. 112).

LEARNING COMPONENTS

Introductory Activity/Engagement

Connection to the Challenge: Begin each day of this lesson by drawing students' attention to the Product Design Challenge and Rubric. It is important that they have some time each day to think about their final presentation.

Science Connection: To introduce the concept of packaging, hold a discussion on UPS, its role in the global shipping industry, and how that is related to marketing and packaging. Show students the UPS video on "Sustainable Packaging" at *www.youtube.com/watch?v=zX5mA3f7Yms*. Have students watch the video a second time and write down any words that they feel are important to discuss. Then, have students form pairs to share their lists of words. Ask several pairs of students to share their lists with the class, and add the words to a class chart. As students share a word, have them define it in their own words. Prompt student to include the terms *efficient, sustainable, environment, sustainable packaging practices, recyclable, reusable, packaging, shipping, sustainable packaging,* and *environmental stewardship.*

Mathematics Class: Ask students to reflect on their activities during the past three days, and have each student make a list on a note card of the different types of packages they have thrown away or recycled. Have students share items from their lists with the class, and create a class chart of student responses. Pair students and have each pair choose at least 10 of the packages on the class chart. Ask student pairs to discuss the characteristics of each type of packaging and record the characteristics of each package they chose in their STEM Research Notebooks. Ask a few student pairs to share their ideas about the characteristics of the packaging they chose with the class. Ask students to provide the mathematical term for the shapes of a variety of packages (e.g., cereal box = rectangular

prism, can of soda = cylinder). Ask students, "How do you think these packages were shipped to the store? How are they organized in the shipping boxes?"

ELA Class: Draw or display a target with a human figure at the center, along with the words, "You are a target in the market." Review the Where's My Stuff? activity from Lesson 1, discussing as a class how items in stores are marketed by placement and packaging. Students might address interests, societal gender marks such as color used for the product or on the packaging, images (e.g., pictures of girls or boys), and so on. Explain that marketers purposely try to interest young buyers in their products.

Ask students to work in pairs to brainstorm ideas about why marketers would target young people even though they rarely have good-paying jobs or a full say in family purchases. Returning to a whole-class setting, invite students to share ideas that they generated. Record these ideas in a displayed document or on a chart to which they may return at a later date.

Ask students to share their own experiences with marketing. Ask students what purchases they have made after seeing products on television, in print, or on social media. Have students complete a STEM Research Notebook prompt reflecting on how they are influenced by marketing.

STEM Research Notebook Prompt

Have students respond to the following questions:

- *Do you think you are influenced by marketing? Why or why not?*
- *Where do you get information about the products you buy?*
- *What type of marketing do you think is most effective? (for example, television advertisements, social media advertisements, billboards or signs, print advertisements, promotional items)*

Activity/Exploration

Social Studies Connection: Not applicable.

Science Connection: Students begin this activity by exploring sustainability. Discuss the term *sustainability* with students. You can have them think about nonpackaging issues where sustainability is important. Share that sustainability means considering what we need to live and how those needs can be met without affecting the ability of people to meet their needs in the future.

Ask students to share their ideas about what sustainability has to do with packaging, creating a class list of student ideas. Ask students if they have ever ordered items from a website, from a television advertisement, or from a mail-order catalog, and then ask them to report how the product was delivered and by what company. Point out to students that as online shopping becomes more common, more and more products are being shipped by organizations such as USPS, FedEx, and UPS. Ask students for their ideas about how the increase in shipping products to consumers' homes might affect sustainability (e.g., using more fuel, using more packaging). Tell students that these organizations each address sustainability in some way.

Assign one shipping organization (USPS, FedEx, and UPS) to each team of three to four students (more than one team will research each organization). For information about each organization's sustainability initiatives see Preparation for Lesson 2 on page 106. Have students respond to the following questions in their STEM Research Notebooks:

- What organization are you researching?
- Does the sustainability initiative have a name? If so, what is it?
- What is the purpose of the initiative?
- What is this organization doing to become more sustainable?
- What more could this organization do to improve sustainability?

STEM Research Notebook Prompt

Ask students to respond to the following prompts in their STEM Research Notebooks, and then discuss as a class:

- *What are some reasons for the use of sustainable packaging?*
- *Why is it important for shippers to encourage their customers to use more sustainable packaging?*
- *Why is it important to shippers to use more sustainable packaging? (Address several issues, such as fuel savings, fewer flights, and fewer trucks on the road.)*

As homework, ask students to observe the packaging of products they use at home. Have students record how much packaging their households threw away in the past week and answer the following questions: How much was recyclable? How much will go into the landfill? If your household disposes of this amount of packaging each week all year long, how many product packages would it dispose of in a year? Based on your household's packaging disposal, estimate the weight of packaging that would go to a landfill in a year. Estimate the weight of the packaging that your household would recycle in a year. Then, write a paragraph that offers ideas about how your household could reduce the amount of packaging it sends to the landfill.

In class the next day, pair students and have them share with their partners how much packaging their households use and how much of this ends up in a landfill. Pose the following question: "How do we know if a package is recyclable?" Group students in twos or threes and assign each group one of the following package types: common plastics, cardboard, coated paper, glass, aluminum, or bioplastic. Then, tell students to use the internet to explore their packaging type, answering the following questions: "What is the life cycle of your packaging type? Discuss what a life cycle is in relation to packaging and sustainability. Is your packaging type sustainable? What is the history of development of this packaging type? How prevalent is its use? What are the dangers associated with this type of packaging? What is its impact on the environment? How can we make this product more sustainable?"

Next, have students create posters that share this information. Assess students' posters using Rubric 6 (p. 120). Have students do a gallery walk near the end of class to examine the other groups' posters. As they walk around, students should take notes in their STEM Research Notebooks about the sustainability of each type of packaging.

STEM Research Notebook Prompt

Have students respond in their STEM Research Notebooks to the following questions, which also connect with the Product Design Challenge, then share their answers with the class using a think-pair-share strategy:

- *If you had to choose a packaging type for your Product Design Challenge, which would you choose and why?*
- *Is there a more sustainable option? Why or why not?*

Mathematics Class: Hide a perspective drawing of a three-dimensional image so students cannot see it. Tell students they will have three seconds to view the object and will have to draw what they saw. (Be sure to give them only three seconds to develop mental images.) After students complete their initial drawings, give them a second look and allow them to refine their drawings.

Then, ask several students to share the answers to the following questions with the class:

- What did you see?
- How did you see it?
- What did you draw first? Why?
- What did you draw next?
- How else did you see it?
- How do you know that is a [right angle, circle, rectangle, rhombus, cube, rectangular prism, etc.]? (This question will vary based on the image you

chose. Students may see the two-dimensional shapes in the drawing of a three-dimensional object.)

- What are the properties of this shape? (Properties include how many sides the shape has, whether opposite sides are parallel to one another, whether opposite sides are equal or unequal in length, and whether all angles are the same.)
- Are there other shapes that have the same properties? Where do you see these shapes in your daily life?

Conduct a three-dimensional shape sort with your students. Provide each team of three or four students with at least six three-dimensional shapes (using real-world objects, premade math manipulatives, and images of three-dimensional shapes). Choose shapes that students will be able to draw from different perspectives. Have student groups each choose a favorite shape and place this shape in the center of the table. Then, ask teams to each find one other shape that is similar and one shape that is different among the shapes you provided or other items in the classroom. Have each team share its shapes with the class, highlighting the similarities and differences.

Then, have student teams place the shapes back in the center of the table and choose one shape at random. Once they have done this, teams should place all the shapes that are similar to this shape next to it. They should place all the shapes that are different in a separate pile. Have each team share with the class why they put the shapes in these categories.

Next, have student teams sort all their shapes into categories. Look for misunderstandings and surface-level responses (e.g., focusing solely on color or texture). Have them sort again by looking at the properties of the shapes. Discuss with students. You may wish to use Venn diagrams and have students look for more than one characteristic at a time. It makes the sort much more challenging and helps students focus on properties of three-dimensional shapes. Next, have each group review another team's sort and identify which property or properties their peers used to sort their shapes.

Choose a few groups to share their sorts with the class. Try to choose a group that sorted by edges and angles and another group that sorted by faces and surfaces. You can use a digital camera to take pictures of the sorts or bring all students to a table to share. If time permits, have students create anchor charts that list the properties of three-dimensional figures. Display these posters around the room. Have students complete a STEM Research Notebook prompt in which they connect their learning about shapes with the module challenge.

STEM Research Notebook Prompt

Have students respond to the following question: *How will what you learned today help you with your design challenge?*

ELA Class: Have students work in pairs to research youth marketing resources. Provide students with a variety of articles and websites about youth marketing to review (see Preparation for Lesson 2, p. 105). Each student pair should review two different resources.

Before students begin their research, hold a class discussion on the concept of audience, making sure students understand that the authors may not be writing to youth as consumers, but instead may be writing to people who want to market to youth, parents or other adults who want to know how marketing could affect youth. Remind students of the benefit of recognizing the structure of a text to support comprehension (refer to the previous lesson on problem-solution text structure). Explain that texts may use a question-and-answer text structure. With the question-and-answer text structure, the author poses a question to get the reader thinking about the topic, then answers that question. Tell students to be alert to this structure and to the use of headings to call reader attention to topics.

Each student pair should answer the following questions in their STEM Research Notebooks for two different resources:

- What organization or individual is providing this information?
- To what audience is this resource directed? (for example, people who want to pursue a career in marketing, the general public, or youth who see this type of marketing)
- What is the purpose of this resource? (for example, to inform people about how to pursue a career in marketing, to show the effects of youth marketing, or to provide instructions about how to market to youth)
- Why are marketers interested in targeting youth?
- What kind of marketing is discussed?
- Does this resource express an opinion about youth marketing? If so, what is the opinion?
- What facts are provided about youth marketing in this resource?
- What facts or ideas were interesting to you?
- What questions do you have after reading this information?

After students complete their research, hold a class discussion, having students share their findings for resources they accessed. Return to the question of why marketers are interested in targeting sales to youth. Ask students what insights they gained about youth marketing from their research. Ask students what else they are now wondering about marketing and how they feel about being the target of marketing efforts.

Social Studies Connection: Students have been exploring demographics in ELA class. Pair students and have them work together to describe what is meant by the term *demographics* and discuss how marketers use demographics to sell products. Have student pairs share their ideas with the class, and hold a class discussion about the meaning of demographics and how marketers use their knowledge of demographics to sell products.

As young consumers themselves, what do they think about being a target of the market? Break students into small groups to discuss this question before conducting a whole-class discussion. Have students from each group share their thoughts about the target market reading, and record responses on a chart or board that will be accessible for future use. Work toward the clear understanding that advertising messages are constructed by people whose goal it is to persuade you to buy their product.

Engage students in a discussion about the geography of marketing, and link this to students' understanding of demographics. Explain that marketing has geographic aspects related to demographics. Marketing uses geographic information in the planning and implementation of marketing activities (product, promotion, price, or place). Provide some examples of how marketers would use different geographic information to make marketing decisions (e.g., "Where would a company market snow shovels, and to whom would they market them?"). Have students explore how the middle school student fits into marketing demographics by exploring the demographics of their class. Working as a class, have students identify several demographic characteristics that marketers might be interested in (e.g., age, gender, hobbies, interests, needs, purchasing power). As a class, develop a demographic survey involving these characteristics. Have each student complete the survey, and then discuss the survey findings as a class.

The following day, have students conduct research to determine where the products they chose for the final challenge originate with respect to where they are shipped. How did the products get where they are now? What are the social and economic implications of where their products were manufactured?

Explanation

Science Connection: Have students complete the STEM Resource Notebook prompt below.

STEM Research Notebook Prompt

As students come into class, have them answer the following question in their STEM Research Notebooks, and then discuss students' responses as a class: *If you were to tell a third grader what it means for a product's packaging to be sustainable, what would you tell him or her? Also explain why it is important for a product's packaging to be sustainable.*

Explain to students the characteristics of sustainable packaging listed in the Teacher Background Information section. Divide students into groups of two or three, and assign each group one of the guidelines of sustainable packaging. Each group should create a PowerPoint slide that explains what the guideline means. Assemble these slides to make a class sustainability presentation (see Rubric 7, p. 121). Share the class PowerPoint presentation and have each group present its slide. Clarify any misconceptions and errors in the presentation. Have students write down new vocabulary from the slide show in their STEM Research Notebooks.

Mathematics Class: Ask students the following question: "If a square has four lines of symmetry, what about cubes?" Explain to students that three-dimensional shapes don't have lines of symmetry; they have planes of symmetry. Pose the question, "How many planes of symmetry does a cube have?" Bring up an interactive website such as *http://illuminations.nctm.org/Activity.aspx?id=4095* or *www.math10.com/en/geometry/geogebra/geogebra.html*. Build a 2 × 2 × 2 cube (be sure to fold the cube up so that students don't see the net just yet), and have students attempt to find all the planes of symmetry for this cube. Ask, "What about a 3 × 3 × 3 cube? What are the properties of this cube? What do you think the cube will look like if it is unfolded?" Unfold for students, then ask, "What two-dimensional shapes do you see? Where do you see this shape in product packaging?" Solicit ideas from students.

Divide students into groups of three or four, and give each group three to five different types of packages that represent a variety of shapes, such as cylinders, cubes, prisms, spheres, and pyramids (some ideas are oatmeal containers, cereal boxes, bottles with labels, cans, shoeboxes, spaghetti boxes, macaroni and cheese boxes, individually wrapped rolls of toilet paper or paper towels, candy containers, takeout containers).

Tell students to sketch the packages in their STEM Research Notebooks and label the various geometric shapes they observe. Then, have teams each choose one of their packages and estimate how much material they would need to create the package, prompting students to consider the area, or size of the surface, of the package. Once students have completed their package exploration, have teams share their findings with the class. Have students describe how they determined how much material would be needed to create these packages. Discuss with students how they estimated or calculated the total area of the surface of a three-dimensional figure. Introduce and define the term *surface area*, and have students write this term and its definition in their STEM Research Notebooks. If students need additional information about volume and surface area, use the websites suggested above or another online resource to review these concepts.

Discuss as a class how surface area might influence shipping of products. Choose a package and ask students, "How do you think these packages would be shipped in bulk to the store?"

Discuss nested packaging with the class, asking, "How many packages of this shape would fit in a box _____ size? What about this shape? Why do you think they ship them this way? How many 1 inch cubes would it take to fill this box? How many 1 cm cubes? How do you know? What would be the volume of your shape?"

ELA Class: Not applicable.

Social Studies Connection: Not Applicable.

Elaboration/Application of Knowledge

Science Connection: Have student teams each create a Life Cycle poster for their chosen module challenge product packaging and develop a plan to improve the sustainability of the packaging (see Rubric 7, p. 121). You may want to have a guest speaker from a local company come in and provide some strategies for students to use more sustainable packaging. Then, have student teams present their posters to the class. After all the presentations, have students brainstorm the materials they want to use for the Product Design Challenge.

Mathematics Class: Tell students to each select five uniquely shaped packages. Then, for each package, sketch the net, determine the surface area, and calculate the volume. Have students develop the formula for finding the surface area and volume of three-dimensional figures. Hold a class discussion about how to find these values for cones.

The next day, as students come into class, have the following problem ready for them to solve: Ace Packaging created a package with a volume of 24 cubic inches. However, in the company's description to the manufacturer, it forgot to say what the package dimensions and shape were. Help the manufacturer out by creating as many packages as you can with this volume. Determine the surface area of your packages as well, so that you know how much material to order. After students have had time to solve the problem, bring them together to share their solutions.

STEM Research Notebook Prompt

Have students respond to the following question in their STEM Research Notebooks, and then discuss their responses as a class: *Why do we need to know about surface area and volume to complete our challenge?*

Then, have students brainstorm the materials they want to use for the Product Design Challenge and draw a scale drawing of their prototype package.

STEM Research Notebook Prompt

Have students respond to the following questions in their STEM Research Notebooks, and then discuss as a class: *If a manufacturer wanted to use nested packaging to ship 24 cubic inch containers to the store, how many would fit in a packing box that measures 10 ft × 4 ft × 2 ft (length × width × height)? Why do we need to know about surface area and volume to complete our challenge?*

An optional extension is to assign one of the demographic characteristics identified in the social studies connection in the Activity/Exploration section (see p. 112) to each team of three or four students, and have them represent the data visually in two ways, such as in a bar graph and a pie chart. Have students share their representations as a class. Choose one representation for each demographic characteristic to display to create a visual demographic overview of the class.

ELA Class: Show an example of a commercial aimed at children. Examples are readily available on the internet. One source is "Toys & Games TV Commercials" at *www.ispot.tv/browse/w.dQ/life-and-entertainment/toys-and-games*, which has links to many such commercials. Have the class discuss whom the commercial seems to be targeting and how the advertisers seem to be trying to appeal to the audience.

Have students view several more child-targeted commercials to look for trends, considering the following questions: What do these commercials have in common? What do you notice about how marketers target even young children? What do you think about marketers targeting young children? If possible, have small groups of students view selected commercials together, then discuss the trends as a team before reporting to the larger group.

It is important that all consumers develop awareness of marketing techniques so they can make informed decisions. Marketing to children raises particular concerns for many people, especially parents. In a blog entry on the Common Sense Media website titled "Sneaky Ways Advertisers Target Kids" (*www.commonsensemedia.org/blog/sneaky-ways-advertisers-target-kids*), author and parent Carolyn Knorr discusses her concern about the potential effects of marketing to children and teens. Show students how this is a good example of argumentative writing, where the author seeks to persuade the reader to take a particular stance and, usually, an action on a particular topic. Have the class read the blog entry from their own perspectives as youth who are marketed to and determine what they think about her argument. Alternatively, you may use a variety of resources available about marketing to children, such as the American Psychological Association's "Report of the APA Task Force on Advertising and Marketing to Children" at *www.apa.org/pubs/info/reports/advertising-children.aspx*, as a foundation for writing a sample argumentative piece for analysis.

Display the Knorr blog entry in front of the class and give a quick minilesson about how the author uses nonfiction text features—boldface print, embedded links, headings, and bullets—to call attention to important information. Depending on the needs of your students, you might review the strategy of turning a heading into a question. In addition, call attention to the teaser at the top of this blog page and discuss briefly how it gives the reader an overview of what will come. Remind readers to always activate prior knowledge before reading to help them understand what they are going to read. In this case, students should think about what they already know about marketing to young people.

Provide each student with a copy of the blog entry discussed above. Have them use sticky notes to mark what they view as particularly important information as they read. After reading the text, hold a discussion about their responses to the article, and record their comments on an accessible chart or board.

Explain to students that they are going use the reading strategies of summarizing and synthesizing to help them craft a response to the parent's blog entry from a middle school student's perspective. Summarizing involves pulling out the most important information and putting it into their own words. Use the recorded discussion responses to demonstrate how students are already doing this. Synthesizing is putting pieces of important information together in a new way to create something more and different from what was read, in this case, a response to the article that brings in the middle school perspective. Display a chart similar to the one in Figure 4.1.

Figure 4.1. Sample Synthesizing Chart

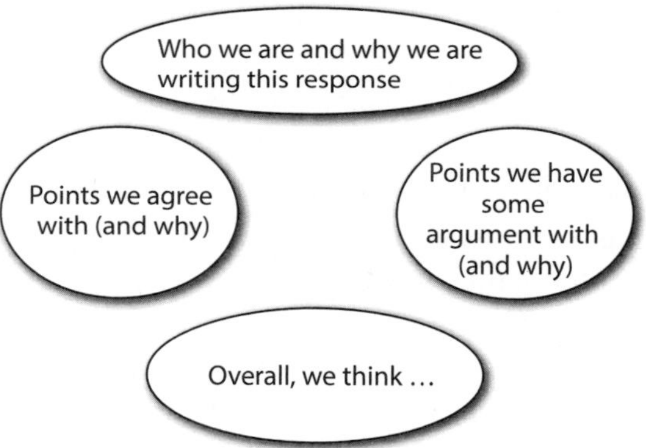

Work with students to identify a simple introduction that reflects who they are and why they are creating the response. Using either virtual sticky notes or actual sticky notes displayed on a document camera or overhead projector, have students identify points with which they agree and points about which they have some argument. Tell students that an argument doesn't have to be complete disagreement but might also be a request for additional information, asking for clarification of an unclear point, or asking another type of question. Record each point students identify on a sticky note and post under the appropriate area of the chart (agreement/argument).

Use this chart as a guide while creating a communally constructed class response to the article. During construction of the response, actively model and engage students in crafting a well-organized and supported argumentative response.

Social Studies Connection: Show students a video that focuses on social justice issues related to the production or packaging of goods, such as "Nike Sweatshops: Behind the Swoosh" at *www.youtube.com/watch?t=1219&v=M5uYCWVfuPQ*. You may want to show the entire video or several clips. After students have watched the video, ask them questions related to the social justice issues raised, such as equality, working conditions, pay, environmental and health impacts of manufacturing waste, and implications for unionizing. Discuss social justice with students and explain that some of the information in the video may be troubling. However, they will be creating an action plan to address the issue as if they were the chief executive officer (CEO) of the company.

Have students research the social justice issues related to their chosen product for the module challenge, and then create a presentation that shows how they would address these injustices if they were the CEO of the company. Then, have students give their social justice presentations to the class (see Rubric 8, p. 122). As time permits, have students develop a commercial that would draw attention to a local social justice issue.

Evaluation/Assessment

Performance Tasks

- Rubric 6: Packaging Type Research Checklist
- Rubric 7: Life Cycle Poster Rubric
- Rubric 8: Social Justice Presentation Checklist

Other Measures

- STEM Research Notebook prompts
- PowerPoint slides
- Participation in class discussions

INTERNET RESOURCES

Sustainability and sustainable packaging

- *https://sustainablepackaging.org/wp-content/uploads/2017/09/Definition-of-Sustainable-Packaging.pdf*
- *www.packworld.com/design/sustainability*

- *www.theguardian.com/sustainable-business/2014/jul/18/good-product-bad-package-plastic-recycle-mistakes*

Manufacturing and outsourcing

- *www.crf-usa.org/bill-of-rights-in-action/bria-17-2-c-globalization-and-worker-rights*
- *https://globalexchange.org/campaigns/legacy-campaigns/fair-trade*
- *www.slideshare.net/anilsural/is-outsourcing-exploitation-or-chance-for-developing-countries*
- *www.theguardian.com/commentisfree/2013/apr/29/bangladesh-factory-tragedy-sweatshop-economics*
- *www.chinalaborwatch.org/report/52*

Teaching about social justice

- *www.reachandteach.com/content/index.php?topic=socialjustice*
- *www.educationworld.com/a_lesson/social-justice-activities-students.shtml*
- *www.weareteachers.com/books-about-social-justice*

"Sustainable Packaging" video

- *www.youtube.com/watch?v=zX5mA3f7Yms*

Youth marketing

- *www.marketing-schools.org/types-of-marketing/youth-marketing.html*
- *www.cbsnews.com/news/resources-marketing-to-kids*
- *https://info.4imprint.com/wp-content/uploads/1P-13-0711-Succeeding.pdf*
- *http://sugarydrinkfacts.org/resources/SugaryDrinkFACTS_ReportSummary.pdf*
- *www.commercialfreechildhood.org/resource/marketing-children-overview*

Shipping organization sustainability initiatives

- *https://about.usps.com/what-we-are-doing/green/welcome.htm*
- *www.fedex.com/cv/about/sustainability/index.html*
- *https://sustainability.ups.com/committed-to-more*

Interactive website on cubes

- *http://illuminations.nctm.org/Activity.aspx?id=4095*

- *www.math10.com/en/geometry/geogebra/geogebra.html*

"Toys & Games TV Commercials"

- *www.ispot.tv/browse/w.dQ/life-and-entertainment/toys-and-games*

"Sneaky Ways Advertisers Target Kids"

- *www.commonsensemedia.org/blog/sneaky-ways-advertisers-target-kids*

"Report of the APA Task Force on Advertising and Marketing to Children"

- *www.apa.org/pubs/info/reports/advertising-children.aspx*

"Nike Sweatshops: Behind the Swoosh" video

- *www.youtube.com/watch?t=1219&v=M5uYCWVfuPQ*

Rubric 6: Packaging Type Research Checklist

Name ________________________________

Criteria	Yes/No	Comments
Students provided accurate information on the life cycle of their packaging.		
Students accurately discussed the sustainability of their packaging.		
Students provided an accurate history of the development of their packaging type.		
Students accurately discussed the prevalence of the use of their packaging type.		
Students accurately discussed the dangers associated with their assigned packaging type.		
Students accurately discussed their assigned packaging's impact on the environment.		
Students discussed how the packaging can be made more sustainable.		
Students worked together on the project as a team.		
The presentation was neat and creative. Information was in a logical sequence and flowed well.		

Rubric 7: Life Cycle Poster Rubric

Name ______________________________

Content	1 Below Mastery	2 Approaching Mastery	3 At Mastery	4 Advanced
SUSTAINABILITY PLAN	The group did not include a sustainability plan in the poster.	The sustainability plan is included but does not reflect an understanding of the need for sustainable packaging.	The sustainability plan was thorough and reflected understanding of the need for sustainable packaging.	The sustainability plan was thorough and innovative and included information presented in the Sustainable Packaging PowerPoint presentation.
LIFE CYCLE	The group did not include a sustainability life cycle or the poster was filled with inaccuracies.	A life cycle was included but was not complete and accurate. Some key information was missing.	The life cycle was included on the poster accurately.	The life cycle was included on the poster accurately and included several interesting details.
ORGANIZATION OF POSTER	The poster was disorganized and had no apparent structure.	The poster was somewhat organized but the structure may not have an effective flow.	The poster was well organized.	The poster was well organized, easy to understand, and effectively drew the reader's attention to important points.
COMMENTS:				

Rubric 8: Social Justice Presentation Checklist		
Name ______		
Criteria	**Yes/No**	**Comments**
Students discussed issues of equality in their presentation.		
Students discussed working conditions in their presentations.		
Students discussed environmental and health impacts of manufacturing waste in their presentations.		
Students discussed implications of unionizing in their presentation.		
Students discussed how they would address social justice issues with the CEO of their product's company.		
Students used graphics appropriately.		
Students worked together on the project as a team.		
The presentation was neat and creative. Information was in a logical sequence and flowed well.		

Lesson Plan 3: Marketing Your Product in a Global Economy

In this lesson, students learn about the marketing profession and world economy. They also continue to explore how marketing influences them as consumers. Students work in groups to design and ultimately present a marketing plan for the team's product. The process includes logo design and the creation of an advertisement. Additionally, students learn about misleading statistics and potential consequences of using statistics that aren't accurate or appropriate for the context. Students complete their Product Design Challenge solutions and present them to a panel of fictional company executives.

ESSENTIAL QUESTIONS

- How do companies create, design, and market products to purposely target a specific audience to maximize profits?
- What are some global issues associated with creating, designing, and marketing products?
- How do marketers use mathematics responsibly?

ESTABLISHED GOALS AND OBJECTIVES

At the conclusion of this lesson, students will be able to do the following:

- Describe that marketing is a complex process that requires feedback from a target market, and revisions are often needed
- Determine whether the statistics in a media campaign are misleading and understand that it is important for the consumer to fact check
- Understand that all media messages are constructed, and when engaging with a media message, consider who created it and for what purpose
- Share how media messages are constructed using a creative language with its own rules, and when engaging with a media message, consider what techniques are being employed to attract buyer attention
- Understand that different people experience the same media message differently, and consider their own impressions and how others might view the message differently
- Understand that media have embedded values and points of view, and consider what values and points of views are represented

- Understand that most media messages are designed to gain profit or power, and consider why a message was sent and how an effective marketing message is created
- Select and use multiple forms of media (visual and textual) to convey information about a product and persuade an audience to buy it

TIME REQUIRED

- 12 days (approximately 45 minutes each day; see Tables 3.8–3.10, pp. 41–43)

MATERIALS

Required Materials for Lesson 3

- STEM Research Notebook
- Computer, tablets, or laptops with internet access for student research
- Projector
- Poster board
- Markers

Additional Materials for Social Studies Connection

- World map or Google Earth

Additional Materials for ELA Class

- Magazine pictures, art supplies, or electronic media images for students to create visual display boards
- Sticky notes

CONTENT STANDARDS AND KEY VOCABULARY

Table 4.7 lists the content standards from the *NGSS, CCSS,* and the Framework for 21st Century Learning that this lesson addresses, and Table 4.8, (p. 130) presents the key vocabulary. Vocabulary terms are provided for both teacher and student use. Teachers may choose to introduce some or all of the terms to students.

Table 4.7. Content Standards Addressed in STEM Road Map Module Lesson 3

NEXT GENERATION SCIENCE STANDARDS

PERFORMANCE EXPECTATIONS

- MS-ETS1-1. Define the criteria and constraints of a design problem with sufficient precision to ensure a successful solution, taking into account relevant scientific principles and potential impacts on people and the natural environment that may limit possible solutions.
- MS-ETS1-2. Evaluate competing design solutions using a systematic process to determine how well they meet the criteria and constraints of the problem.
- MS-ETS1-3. Analyze data from tests to determine similarities and differences among several design solutions to identify the best characteristics of each that can be combined into a new solution to better meet the criteria for success.

DISCIPLINARY CORE IDEAS

ETS1.A. Defining and Delimiting Engineering Problems

- The more precisely a design task's criteria and constraints can be defined, the more likely it is that the designed solution will be successful. Specification of constraints includes consideration of scientific principles and other relevant knowledge that are likely to limit possible solutions.

ETS1.B. Developing Possible Solutions

- There are systematic processes for evaluating solutions with respect to how well they meet the criteria and constraints of a problem.
- Sometimes parts of different solutions can be combined to create a solution that is better than any of its predecessors.

ETS1.C. Optimizing the Design Solution

- Although one design may not perform the best across all tests, identifying the characteristics of the design that performed the best in each test can provide useful information for the redesign process—that is, some of those characteristics may be incorporated into the new design.

CROSSCUTTING CONCEPTS

Influence of Science, Engineering and Technology on Society and the Natural World

- All human activity draws on natural resources and has both short- and long-term consequences, positive as well as negative, for the health of people and the natural environment.
- The uses of technologies and any limitations on their use are driven by individual or societal needs, desires, and values; by the findings of scientific research; and by differences in such factors as climate, natural resources, and economic conditions. Thus technology use varies from region to region and over time.

Continued

Table 4.7. (*continued*)

Cause and Effect

- Relationships can be classified as causal or correlational, and correlation does not necessarily imply causation.

Science Is a Human Endeavor

- Scientists and engineers are guided by habits of mind such as intellectual honesty, tolerance of ambiguity, skepticism, and openness to new ideas.

SCIENCE AND ENGINEERING PRACTICES

Asking Questions and Defining Problems

- Asking questions and defining problems in grades 6–8 builds on grades K–5 experiences and progresses to specifying relationships between variables, and clarifying arguments and models.
- Define a design problem that can be solved through the development of an object, tool, process or system and includes multiple criteria and constraints, including scientific knowledge that may limit possible solutions.

Obtaining, Evaluating, and Communicating Information

- Obtaining, evaluating, and communicating information in 6–8 builds on K–5 and progresses to evaluating the merit and validity of ideas and methods.
- Gather, read, and synthesize information from multiple appropriate sources and assess the credibility, accuracy, and possible bias of each publication and methods used, and describe how they are supported or not supported by evidence.

Planning and Carrying Out Investigations

- Planning and carrying out investigations in 6–8 builds on K–5 experiences and progresses to include investigations that use multiple variables and provide evidence to support explanations or solutions.
- Conduct an investigation to produce data to serve as the basis for evidence that meet the goals of an investigation.

Analyzing and Interpreting Data

- Analyzing data in 6–8 builds on K–5 experiences and progresses to extending quantitative analysis to investigations, distinguishing between correlation and causation, and basic statistical techniques of data and error analysis.
- Analyze and interpret data to determine similarities and differences in findings.

Constructing Explanations and Designing Solutions

- Constructing explanations and designing solutions in 6–8 builds on K–5 experiences and progresses to include constructing explanations and designing solutions supported by multiple sources of evidence consistent with scientific ideas, principles, and theories.
- Apply scientific principles to design an object, tool, process or system.

Continued

Table 4.7. (*continued*)

Engaging in Argument From Evidence

- Engaging in argument from evidence in 6–8 builds on K–5 experiences and progresses to constructing a convincing argument that supports or refutes claims for either explanations or solutions about the natural and designed world(s).
- Use an oral and written argument supported by evidence to support or refute an explanation or a model for a phenomenon.
- Evaluate competing design solutions based on jointly developed and agreed-upon design criteria.

COMMON CORE STATE STANDARDS FOR MATHEMATICS

MATHEMATICAL PRACTICES

- MP1. Make sense of problems and persevere in solving them.
- MP3. Construct viable arguments and critique the reasoning of others.
- MP4. Model with mathematics.
- MP5. Use appropriate tools strategically.
- MP6. Attend to precision.

MATHEMATICAL CONTENT

- 6.RP.A.2. Understand the concept of a unit rate a/b associated with a ratio $a:b$ with $b \neq 0$, and use rate language in the context of a ratio relationship.
- 6.RP.A.3. Find a percent of a quantity as a rate per 100 (e.g., 30% of a quantity means 30/100 times the quantity); solve problems involving finding the whole, given a part and the percent.
- 6.EEA.2. Write, read, and evaluate expressions in which letters stand for numbers.
- 6.EEB.6. Use variables to represent numbers and write expressions when solving a real-world or mathematical problem; understand that a variable can represent an unknown number, or, depending on the purpose at hand, any number in a specified set.
- 6.SP.A.1. Recognize a statistical question as one that anticipates variability in the data related to the question and accounts for it in the answers.
- 6.SP.A.3. Recognize that a measure of center for a numerical data set summarizes all of its values with a single number, while a measure of variation describes how its values vary with a single number.
- 6.SP.B.5. Describing the nature of the attribute under investigation, including how it was measured and its units of measurement.

Continued

Table 4.7. (*continued*)

COMMON CORE STATE STANDARDS FOR ENGLISH LANGUAGE ARTS

READING STANDARDS

- RH.6-8.4. Determine the meaning of words and phrases as they are used in a text, including vocabulary specific to domains related to history/social studies.
- RI.6.2. Determine a central idea of a text and how it is conveyed through particular details; provide a summary of the text distinct from personal opinions or judgments.
- RI.6.6. Determine an author's point of view or purpose in a text and explain how it is conveyed in the text.

WRITING STANDARDS

- W.6.4. Produce clear and coherent writing in which the development, organization, and style are appropriate to task, purpose, and audience.
- W.6.6. Use technology, including the Internet, to produce and publish writing as well as to interact and collaborate with others; demonstrate sufficient command of keyboarding skills to type a minimum of three pages in a single sitting.
- WHST.6-8.4. Produce clear and coherent writing in which the development, organization, and style are appropriate to task, purpose, and audience.
- WHST.6-8.6. Use technology, including the Internet, to produce and publish writing and present the relationships between information and ideas clearly and efficiently.
- WHST.6-8.8. Gather relevant information from multiple print and digital sources, using search terms effectively; assess the credibility and accuracy of each source; and quote or paraphrase the data and conclusions of others while avoiding plagiarism and following a standard format for citation.
- WHST.6.10. Write routinely over extended time frames (time for reflection and revision) and shorter time frames (a single sitting or a day or two) for a range of discipline-specific tasks, purposes, and audiences.

FRAMEWORK FOR 21ST CENTURY LEARNING

21st Century Themes

- Environmental Literacy
- Health Literacy
- Global Awareness
- Financial, Economic, Business, and Entrepreneurial Literacy
- Civic Literacy

Continued

Table 4.7. (*continued*)

Learning and Innovation Skills

- Creativity and Innovation
- Critical Thinking and Problem Solving
- Communication and Collaboration

Information, Media and Technology Skills

- Information Literacy
- Media Literacy
- ICT Literacy

Life and Career Skills

- Flexibility and Adaptability
- Initiative and Self-Direction
- Social and Cross Cultural Skills
- Accountability, Leadership, and Responsibility

Table 4.8. Key Vocabulary in Lesson 3

Key Vocabulary	Definition
average order	the mean number of items ordered from a company
constrict	become narrow
correlation	a relationship or connection between two or more things
DNA	deoxyribonucleic acid; material that carries genetic information in nearly all living things
economic downturn	a drop in the growth rate of an economy, which can occur at a business, country, regional, or global level
economic environment	the combination of economic factors that together influence the buying behavior of people or institutions
economic interdependence	globally, countries' reliance on each other to provide products and services, trading with each other through imports and exports
economy	the system by which a country or region's products and services are produced, sold, and used
emerging market	a less developed country whose economy is becoming more advanced
exports	goods or services sent to other countries or regions for sale
fair trade	a system to ensure that fair prices are paid to the people who make products, especially to aid those in less developed countries
free trade	a system of buying and selling goods between countries with no restrictions or taxes on imports
global	involving or affecting the entire world; worldwide
global economy	the system of industry and trade among the interrelated economies of countries around the world, considered as a single worldwide economy in which countries can affect the others negatively or positively
gross profit	total revenue minus the cost of making and selling a product
imports	goods or services brought in from other countries or regions for sale
investment	the purchase of an item for future use or to generate income in the future
marketing costs	the amount of money spent to market a product
misleading statistics	statistics that are used incorrectly or to misinform people

Continued

Table 4.8. (*continued*)

Key Vocabulary	Definition
needs	goods or resources people must have to survive and thrive
organism	a living thing, such as an animal, plant, or single-celled life form, that engages in life processes such as using energy, moving, growing, and reproducing
positive association	a good feeling one has when seeing, hearing, tasting, or feeling a familiar product, image, or song
price	the amount of money one has to pay for a product or service
product	something that is made for sale
replicate	make a copy of itself; reproduce
retention rate	in marketing, the percentage of people who became customers from a potential pool of customers, or the percentage of customers who continue over time to buy a product or use a service being marketed
return on investment (ROI)	the amount of profit from an investment, measured by calculating the difference between the profit and the original investment
revenue	the amount of money a company regularly takes in; income
statistics	a branch of mathematics that organizes and analyzes collections of numerical data about a topic or a population
target market	the particular group of customers who are most likely to be interested in a product
virus	a disease-causing agent, too tiny to be seen with an ordinary microscope, that invades living cells and multiplies there
wants	things that are desired or demanded but are not necessities

TEACHER BACKGROUND INFORMATION

This lesson provides students with the opportunity to develop a marketing plan and advertisement for their product. In this lesson, they explore social issues surrounding marketing and advertisement. They learn about the consequences of false advertisement and how they can avoid being misled by statistics. This lesson focuses on critical understanding and effective production of a variety of texts, including media marketing texts. Students develop understanding of how to read nonfiction/informational and persuasive texts, including using nonfiction text structures and features and reading comprehension strategies. Students develop the media literacy skills necessary to be critical consumers and producers of media messages. Students also wrap up their Product Design Challenge and give presentations to a panel of fictional company executives (e.g., school administrators and community members invited to provide feedback).

How One Catches a Cold

Cold weather does not cause one to catch a cold. Most cold symptoms are caused by the body reacting to a virus, most commonly rhinoviruses. However, recent research shows that exposure to the cold may weaken the immune system and thus may reduce the body's ability to fight rhinovirus infections. The research also suggests that rhinoviruses replicate more easily in cooler temperatures. For more information, see the following websites:

- *www.everydayhealth.com/cold-and-flu/colds-and-the-weather.aspx*
- *www.medicalnewstoday.com/articles/166606.php*
- *www.webmd.com/cold-and-flu/news/20150106/researchers-probe-why-colds-are-more-likely-in-winter*
- *www.pbs.org/wgbh/nova/next/body/scientists-finally-prove-cold-weather-makes-sick*

The Nature of Science

While science shows us that being cold doesn't cause colds, research does show us why we catch more colds in the winter. The myth about the cold giving you a cold was developed as people tried to explain why they caught colds. The science at the time may have suggested that colds were caused by the cold, but current research shows otherwise. At one point, scientists believed that the Sun revolved around Earth as well. This was also disproven by new evidence.

As our knowledge grows and as technology makes it easier for us to explore things on a micro and macro scale, we learn more about the world we live in. The National Science Teachers Association (NSTA) wants science teachers to understand that "science is characterized by the systematic gathering of information through various forms of direct and indirect observations and the testing of this information by methods including, but not limited to, experimentation. The principal product of science is knowledge in the form of naturalistic concepts and the laws and theories related to those concepts" (2000). It also wants students to understand the nature of science. Julie Angle, an associate professor in science education at Oklahoma State University and expert in the nature of science describes four aspects of the nature of science that students must understand as they leave their K–12 experiences: (1) science has its limits, (2) scientific knowledge is inherently uncertain, (3) science can be done poorly and can be misused, and (4) science is culturally and politically influenced (personal communication, September 5, 2015).

Therefore, when discussing science concepts with your students, you should discuss the nature of science as well.

False Advertising and Misleading Statistics

According to Gandhi (2013), the term "snake oil salesman" originated in the late 1800s. Chinese workers who came to the United States to work on the Transcontinental Railroad as indentured laborers used an oil made from the fat of the Chinese water snake to relieve pain associated with arthritis. They shared it with their American coworkers, who found it effective at relieving their pain after a hard day working on the railroad. Clark Stanley, an entrepreneur, decided to try to mimic the Chinese water snake oil by using rattlesnake oil instead. He claimed that he had learned about it from Hopi medicine men and sold it at the 1893 World's Exposition in Chicago, where he actually took a live snake and cut it open to make his liniment in front of a crowd. As it turned out, the liniment Stanley was peddling did not even contain any snake oil. In 1917, authorities discovered that it was made of mineral oil, an oil that was likely beef fat, red pepper, and turpentine. Thus the term "snake oil salesman" became associated with fraud.

The Federal Trade Commission (FTC) was established to enforce provisions of the Federal Trade Commission Act, designed to protect consumers from "unfair or deceptive acts or practices in or affecting commerce" (15 U.S.C. § 45(a)(1)). To learn more about the act and the FTC, visit *www.ftc.gov/about-ftc/what-we-do*. When the FTC finds that a company's claim is false, it files action in federal district court. If a company is found guilty of violating laws, it is usually required to stop the marketing campaign and change the product labels. Companies also are often fined large sums of money and required to compensate the consumers who have been defrauded (through refunds or cash settlements). For more information on the FTC's protection of consumers from false advertising, see "Truth in Advertising" at *www.ftc.gov/news-events/media-resources/truth-advertising*.

It is important for students to understand the implications of false advertising and misleading practices so that they avoid such things in their advertising campaign. To accomplish this, students conduct a case study on Airborne cold remedy. In the commercial, made by the former owners of Airborne, the advertisers mention that they used a double-blind, placebo-controlled study to determine whether the Airborne product prevented colds. According to Dellwo (2017), "a double-blind, placebo-controlled clinical trial is a medical study involving human participants in which neither side knows who's getting what treatment and a placebo is given to a control group." This is considered the gold standard in drug research. The FTC found that the claims made by the former owners had no credible basis, and the company eventually settled out of court. In the settlement, the company agreed to not make any claims that its product would prevent colds. It was also fined $30 million as part of the settlement (FTC 2008).

There are other important cases in recent history that demonstrate to students how companies are punished for false advertising and deceptive practices. Here are links to some of the more well-known cases:

- *www.ftc.gov/news-events/press-releases/2010/12/dannon-agrees-drop-exaggerated-health-claims-activia-yogurt*
- *www.businessinsider.com/false-advertising-scandals-2011-9*
- *www.huffingtonpost.com/2011/01/25/taco-bell-beef-lawsuit_n_813185.html*
- *www.huffingtonpost.com/2014/05/16/companies-lied-to-you_n_5318940.html*
- *www.huffingtonpost.com/2012/08/29/new-balance-toning-shoe-settlement_n_1839537.html*

Students also will look at ways that advertising can use statistics to fool them. To learn more about misleading statistics and how they can influence decision making, see the following resources:

- *http://ed.ted.com/lessons/how-statistics-can-be-misleading-mark-liddell*
- *http://faculty.atu.edu/mfinan/2043/section31.pdf*
- *www.cracked.com/photoplasty_2052_19-ways-you-can-make-statistics-say-whatever-you-want*
- *www.truthpizza.org/logic/stats.htm*
- *www.econoclass.com/misleadingstats.html*

Marketing Return on Investment

Marketers use mathematics to determine their return on investment. Basically, they calculate whether or not their advertisement campaign is increasing profit. For more information on return on investment and other marketing metrics, explore the following sites:

- *www.marketingmo.com/campaigns-execution/how-to-calculate-roi-return-on-investment*
- *www.investopedia.com/articles/personal-finance/053015/how-calculate-roi-marketing-campaign.asp*
- *http://blog.hubspot.com/marketing/math-behind-common-marketing-metrics*

Marketing and Media Literacy

You may have a great product to sell, but if people don't know about it they won't buy it. Marketing is what companies do to let potential buyers, or consumers, know about a

product and entice them to purchase the product. Marketing usually involves advertising, getting a message about the product out to potential buyers. We see and hear advertisements around us every day, during television shows, between articles in a magazine, on roadside billboards, in pop-ups on websites, and more. Marketing involves more than advertising, though. Developing attractive packaging and an appealing logo that consumers will recognize immediately (think McDonald's arches or Nike's Swoosh), pricing products competitively, providing good customer service, and supporting the community through donations and sponsorships are other common marketing strategies that raise consumer awareness of products and encourage them to buy.

To market a product effectively, marketers need to get the word out to as many potential buyers as possible, so they send their messages through media. Media is the plural of medium, a way of getting something done; in this case, media are ways of communicating messages from a source to an audience. Marketers who want to send messages about their products often take advantage of mass media, media that reach big numbers and wide ranges of buyers. Television is a great example of mass media, and the commercials that viewers watch are a form of media message sent through mass media. The goal of commercials is to get people to buy the product; therefore, commercials are a form of mass marketing.

Marketers have one goal: getting you to buy their message. Rarely are messages simple and direct: "This is a good product and you should buy it." Rather, marketers think about who might buy their products and why, and they try to send a message that will make those people in their target market really want to make that purchase. More than just informing us about products, marketing can (and often does) change the way we think about what we want and need and, in the process, even the way we think about who we are and how we should be. Have you ever been around a child watching toy commercials during a television show? Have you ever felt that you needed a certain brand of clothes or shoes to fit in or be cool? The marketers work to make these products seem highly appealing to their target market, the child or teen who can directly influence the buying decisions of the parent. A child's reaction to a commercial is usually immediate and positive toward the product, a reaction that involves no consideration of how well the product might actually work, whether it is needed or appropriate, or its cost. The message has the most power when it is taken in directly and without critical consideration.

Consider all the forms of media with which we engage daily—television, videos, social media platforms, and print texts. All of these are avenues for complex messages that can change the way we think about who we are and what we need and want. Because media messages can be so powerful, and because we want to be in control of our own decisions rather than be manipulated by someone else, we need to be aware of the ways media messages affect us. Media literacy refers to our ability to see beyond the surface of the message to be aware of how and why the message was created and might influence us.

From a marketing standpoint, when we are media literate, we are more likely to make good decisions about what to buy.

To read a media message effectively, thinking for ourselves instead of just buying what we are told, we need to consider who authored the message and what the author is trying to accomplish. We need to be able to recognize techniques that the message author uses to get us to respond in certain ways, and we need to consider the values that come across in the message as well as what values are not represented. We also need to understand that different people will understand the same message differently, considering who might be affected and in what ways by the message.

Media messages have profound effects on individuals, but they also shape society on a larger scale. Marketing can change social perceptions of needs and wants, potentially contributing to social inequity and environmental harm. When those who have resources clamor for products that they come to believe they really need (rather than just want), less of what people really need is produced and shared, so those who have fewer resources may miss out on what they need. Both production effects (such as pollution) and disposal effects (such as filling up landfills with discarded stuff) change the planet for the worse. Because marketing messages are so powerful, to be responsible and healthy consumers, we need to understand that power and be able to harness it for best use.

To assist your students with their marketing campaigns and presentations, see "7 Ways to Create a Successful Integrated Marketing Campaign" at *www.cio.com/article/2377257/online-marketing/7-ways-to-create-a-successful-integrated-marketing-campaign.html* and "Four Secrets to Giving a Great Marketing Presentation" at *www.entrepreneur.com/article/234832*. For more information about media literacy, including teaching suggestions, visit the Center for Media Literacy website at *www.medialit.org*.

Cocoa Production in Ghana

In social studies, students will explore issues associated with the production, shipping, and purchasing of cocoa from Ghana. To teach this lesson, it is important to learn about the history, culture, geography, and economy of Ghana and the role that cocoa plays in everyday life. Ghana is the second-largest exporter of cocoa in the world, after Côte d'Ivoire (WorldAtlas 2017). There are also some social justice issues that you should be aware of in relation to the cocoa trade, including fair trade and child labor. The following websites provide information on the history, culture, and people of Ghana and social justice issues surrounding the cocoa trade:

- *www.britannica.com/place/Ghana*
- *www.ghanaweb.com/GhanaHomePage/history*
- *www.infoplease.com/country/ghana.html*

- *http://crawfurd.dk/africa/ghana_timeline.htm*
- *www.ancient.eu/Ghana*
- *http://thegrio.com/2013/11/02/why-ghana-is-fast-becoming-a-hub-for-african-americans*
- *www.bbc.com/news/world-africa-34936103*
- *www.foodispower.org/slavery-chocolate*
- *http://equalexchange.coop/products/chocolate/faqs/what-makes-fair-trade-cocoa-and-chocolate-products-different*

COMMON MISCONCEPTIONS

Students will have various types of prior knowledge about the concepts introduced in this lesson. Table 4.9 outlines some common misconceptions students may have concerning these concepts. Because of the breadth of students' experiences, it is not possible to anticipate every misconception that students may bring as they approach this lesson. Incorrect or inaccurate prior understanding of concepts can influence student learning in the future, however, so it is important to be alert to misconceptions such as those presented in the table.

Table 4.9. Common Misconceptions About the Concepts in Lesson 3

Topic	Student Misconception	Explanation
Viruses	Antibiotics can be used to treat illnesses caused by viruses.	Antibiotics are effective against illnesses caused by certain bacteria. Viruses are not bacteria, however, and therefore antibiotics are ineffective in treating illnesses caused by viruses. Whereas bacteria are single-celled organisms that can reproduce and exhibit the characteristics of living things, viruses do not have cells and require a host cell to reproduce.
Catching a cold	You can catch a cold from being cold.	The only way to contract the common cold is to come into contact with viruses, such as rhinoviruses and other respiratory viruses, that cause colds.

PREPARATION FOR LESSON 3

Review the Teacher Background Information provided (p. 131), assemble the materials for the lesson, and preview the videos recommended in the Learning Components section below. Contact individuals to act as the panel of company executives for student presentations. These may be school administrators, parents, older students, or other community members. Provide these individuals with an overview of the module challenge and copy of the Product Design Challenge rubric. Members of the panel should be prepared to provide students with feedback on their designs. You may wish to create a feedback form for panel members to use that includes prompts (for example, "what I like about this design," "what I think could be improved about this design," "what questions I have").

You should also have a crush test station available for testing teams' packaging designs. The crush test is the ability of a package to withstand a five pound weight placed on top of it. The station can be set up as described for the Save the Chips Competition (see p. 63).

For the science connection, search online for additional videos, apps, or animations that discuss and show whether being cold makes you sick, as well as animations of how cold viruses function and replicate. Student teams will research whether cold weather causes colds. Ensure that students can access internet resources about the cause of colds or prepare printed materials for student research.

For mathematics class, search for misleading ads in newspapers and magazines, as well as online. Focus specifically on ads that use misleading statistics.

For ELA class, search YouTube for commercials that demonstrate targeted appeal to varied demographics. For instance, Subaru markets to adults with children, focusing on safety and reliability, while Gatorade aims at the young and active demographic. Make sure you collect a wide range, from the commercials targeting children and mothers that are shown during young children's TV shows to the commercials that target older adults who have finished raising their families. Aim for a variety of perspectives, including different genders and ethnicities. Ultimately, you should choose five to seven commercials that represent marketing to a variety of individuals. Also search the internet for product logos for students to examine. There are a number of websites that show a wide range of logos. An interesting one that includes not only a visual overview of logos but also some discussion about the meaning of each is "30 Famous Logos That Have a Hidden Message" at *http://diply.com/famous-logos-hidden-messages?publisher=what-facts&config=20*. Websites with tips on designing logos, such as "Vital Tips for Effective Logo Design" at *www.smashingmagazine.com/2009/08/vital-tips-for-effective-logo-design*, may also be helpful. Be sure to preview sites and consider the appropriateness of the products and advertising for your students.

Also for ELA class, determine how you will display the five core media literacy concepts and five key questions provided at *www.medialit.org/reading-room/*

five-key-questions-form-foundation-media-inquiry so that you can provide students with continuous access to this information. Select some commercial scripts from the "Voice Over Sample Scripts" at *www.voices.com/blog/voice-over-sample-scripts* and determine how you will share those scripts with students. Develop a guide to resources that students can use in creating and communicating their marketing campaigns, and plan to support skills (e.g., media use, production, long-term project planning) as needed. These needs may vary widely from school site to school site, depending on access to media and students' prior knowledge. Whatever the setting, make sure you provide focused, need-based support for the processes involved.

For social studies, research various countries that export cocoa and find out what these countries have in common. Select a variety of countries for your students to explore. Also research modern examples of issues associated with economic interdependence. Specifically, find examples of economic downturns in the United States and other countries that have had an impact on the world economy.

LEARNING COMPONENTS

Introductory Activity/Engagement

Connection to the Challenge: This lesson allows students to develop, test, and present about their product. It is important to give students time each day to do a status update and to revisit the Product Design Challenge Guidelines and Rubric that are in their STEM Research Notebooks.

On the last two days of the module, have students present their products, packaging designs, and campaigns to the panel of "company executives" and subject their product packaging to a crush test.

Science Connection: Have students complete the following activity.

STEM Research Notebook Prompt

Ask students to respond to the following STEM Research Notebook prompt: *You want to go outside to play, but your parents tell you it is too cold to go outside, saying, "You will catch a cold." Does being cold make you catch a cold? Why or why not?*

Mathematics Class: Students will explore misleading statistics in the next mathematics class, and it is important that they learn about real-world examples of misleading statistics and advertising as background. Start off this lesson by asking students if they have ever heard of a "snake oil salesman" and know how the name got its meaning. Then, provide students with an example of a "snake oil salesmen." Video clips and online articles can be found using the search terms *snake oil salesman*. One example is a clip from the video titled "Pete's Dragon Passamaquaddy," at *www.youtube.com/watch?v=QHboMLW-Zn0.*

Hold a class discussion, asking questions such as: Do you think the claims these people made are true? What evidence do they provide that their products work? How do we know whether he is telling the truth? Next, show students a video or article about these snake oil salesmen in history. One example is "The Story of the Snake Oil Salesman" at *www.youtube.com/watch?v=titzrDTfp70*.

Explain that the Federal Trade Commission (FTC) enforces laws that prohibit advertisers and marketers from lying about products. Tell your students more details about what the FTC does based on what you learned from the Teacher Background Information section (see p. 133).

ELA Class: Lead a brief discussion about what it means to be literate. Most students will define literacy as the ability to read and write. Expand on that notion by pointing out that we read more than just printed words and send messages in ways other than by writing. Introduce the term *media literacy* and introduce media literacy concepts, making connections to the work students have already done. Tell students they will learn more about media marketing, using what they know to understand the messages they encounter and to create their own marketing messages.

Show students a television commercial that targets a particular audience, examining together the marketing messages using media literacy questions from the Center for Media Literacy. After the initial discussion, divide students into groups and assign each an additional commercial to analyze using the media literacy questions. Assign groups the advertisements you identified that represent marketing to a variety of audiences (see Preparation for Lesson 3, p. 138). Have each group share its findings, discussing what they noticed about the advertisements and hold a class discussion about how the advertisements were different depending on the audience that was targeted.

Social Studies Connection: Hold a class discussion about what students have learned about marketing, asking students, "What is marketing and why should we learn about it?" Have students share with the class what they have learned about marketing.

Activity/Exploration

Science Connection: Ask students to share their ideas from the STEM Research Notebook prompt they completed in the Introductory Activity/Engagement (see p. 139). Have students who think they will catch a cold from being cold move to one side of the room. Have students who believe that being cold doesn't make them catch a cold move to the other side.

Form teams of students who believe that cold weather causes colds and teams of students who believe that cold weather does not cause colds. Tell students that their task is to find evidence to support their belief that cold temperatures will or will not make people more susceptible to catching a cold. Student teams should find enough evidence to support the claim and present the argument to their peers. Their peers will try to find holes in the argument, so students should make sure that they are well prepared to defend their position. Each team of students should identify several points in support of their position and record them in their STEM Research Notebooks.

The next day, give students five minutes to tighten up their evidence and decide how to present their points in the class debate. Then, pair teams of opposing views and bring teams of students up front to present their arguments and evidence to the class. Have students look for holes in the other team's arguments. Tell them to write down questions they may have and points they want to make as their peers are presenting, and then allow students to ask questions after each group has presented their findings. Was anyone swayed by the other side's arguments?

Show students an animation or video that discusses the issue of whether cold temperatures make you sick. You can find videos using search terms such as "does cold weather cause colds." An example of a video is "Does Being Cold Make You Sick?" at *www.youtube.com/watch?t=10&v=RWiOhlqEDz4*. After watching the video, ask students to indicate, by a show of hands, how many of them have caught a cold after being outside on a cold day. Introduce the term *correlation* as a relationship between two things. Ask students to share their ideas about whether their class vote showed a correlation between cold weather and illness. Introduce the concept of *causation*, which refers to one occurrence being caused by another. Ask students for their ideas about whether correlation is the same as causation, guiding students to understand that correlation is different than causation. You may wish to use some examples of spurious correlations, such as pointing out that ice cream sales are correlated with cases of sunburn (for examples of spurious correlations, see *www.bbc.com/news/magazine-27537142*).

Ask students, "What is a virus?" Ask students to share their ideas about how a virus enters their bodies and makes them sick. Show students an animation of how a flu or cold virus spreads in our body and how our immune system works to protect us from the virus. For example:

- "Flu Attack! How a Virus Invades Your Body" *www.youtube.com/watch?t=206&v=Rpj0emEGShQ*
- "Influenza Animation—Flu Virus Mechanism" *www.youtube.com/watch?v=YSgkoldBNkI*
- "Viruses: Virus Replication and the Mysterious Common Cold" *www.youtube.com/watch?v=PHp6iYDi9ko*

Ask students for their ideas about whether a virus is living or nonliving and how they could determine that. Create a class list of the characteristics of living things: They are made of cells; use resources for energy (have the capacity to take in, transform, and use

energy from the environment); respond to stimuli from the environment; have the capacity to reproduce; and grow and develop. Hold a class discussion about whether viruses meet the criteria in this list. Tell students that for many years viruses were classified as nonliving things; however, as scientists learn more about viruses some believe that they should be considered living things since they have some of the biological equipment needed to reproduce. While this is an ongoing debate among scientists, students should understand that viruses are fundamentally different from bacteria, which are considered living things and can be treated with antibiotics.

Hold a class discussion about viruses and illness, using information from the videos the class viewed and asking questions such as the following:

- What are reasons that we catch the common cold?
- What are some ways that our body fights viruses that can cause a cold?
- What can we do to prevent our bodies from catching a cold?

STEM Research Notebook Prompt

Have students respond to the following prompt in their STEM Research Notebooks, and then discuss as a class: *How can marketers use people's misconceptions about how one catches a cold to encourage people to purchase their products? Can you think of any examples?*

The next day, remind students that they looked at a common myth about how we catch the common cold and that they used science to agree or disagree about whether cold temperatures make us catch colds. Help students think about the nature of science by discussing the four aspects given in the Teacher Background Information section (see p. 132).

Mathematics Class: Pose the following problem to the class: "A recent study found that 5% to 20% of Americans catch a cold each winter. If there are 300 million people in the United States, how many people get a cold each winter?" Give students time to calculate their response, and then have them share their findings with the class. Ask them, "What remedies are out there to help prevent the common cold?"

Show students the Airborne commercial at *www.youtube.com/watch?v=pDoG3gvE8Nc*. What was this product's message? Do you think the product was intended as a cure or a prevention method for the common cold? Why do you think the commercial took place on an airplane?

Have students read the article titled "Does Airborne Really Stave Off Colds?" at *http://abcnews.go.com/GMA/OnCall/story?id=1664514&page=1*. Ask students the following questions (refer to the Teacher Background Information section on p. 133 for details on the Airborne case to help you discuss this with students):

- Do you think that there was some validity to the claims?
- What were the issues with the study that Airborne used?
- What is a double-blind, placebo-controlled study?
- What were the consequences for the company that sold Airborne?
- How did the company address the issue in its packaging? (Explain to students that the product is no longer owned by the company that got in trouble for false advertising, and the packaging has since changed under new ownership.)
- How can you avoid making a similar mistake?

Next, have students research advertisements for their chosen products focusing on the statistics used in the advertisements. Have students record their findings to the following questions in their STEM Research Notebooks:

- Are there any statistics in your product's advertisements?
- If so, are they misleading? If not, what statistics do you think the company could have provided to make its advertisements more powerful?
- How will you use statistics in your advertising campaign?

ELA Class: Connect group findings from the previous day to the media literacy concepts, making sure students understand the various concepts. Put students in the same working groups they had for selecting their products in the science connection. Have students consider the ideal consumers for their products. Using magazine pictures, art supplies, or electronic media images, have students create visual display boards to define their target markets. To accompany their displays, students should each write a brief paragraph describing their products. (Review descriptive writing with students as needed.)

Then, post the students' target market boards and descriptions around the room. Over the next few days, encourage students to attach suggestions written on sticky notes to each other's display boards about ways to reach particular target markets through advertising.

Social Studies Connection: Students will complete a case study of Coca-Cola's marketing strategies. This will help students understand effective marketing strategies and the global reach of marketing that allows a business to have customers in many parts of the world. Have students watch a video about Coca-Cola marketing. A search of YouTube using terms such as *Coca-Cola* and *marketing* will provide videos such as "The Secret Behind Coca-Cola Marketing Strategy" at *www.youtube.com/watch?v=XhMVWzVXNNk.* Provide students with questions before they watch the video so that they can watch for the answers. Questions can include the following:

- What makes Coca-Cola such a recognizable brand all over the world?
- What were some of the effective marketing strategies used by Coca-Cola? Why were these so effective?
- What were some of the ineffective marketing strategies used by Coca-Cola? Why were they ineffective?
- How could you use this information to help you design a stronger marketing campaign for your Product Design Challenge?
- How have these marketing strategies had an impact on you as a target of marketing?

The following day, have a discussion with students on the differences between needs and wants. Make connections to the unequal distribution of resources. What do villages, towns, and nations have to do when they do not have the resources to support their people? How does marketing influence what we consider our needs and our wants?

Explanation

Science Connection: Not Applicable.

Mathematics Class: Tell students that sharing and using false and misleading statistics in marketing can be damaging to the company's reputation and costly. Show students a video on misleading statistics such as "Don't Be Fooled by Bad Statistics" at *www.youtube.com/watch?v=jguYUbcIv8c*. (*Note:* You may want to stop the video at 2:41 and fast forward to 3:08, because the narrator mentions that Santa isn't real, and that may upset some children who still believe in Santa.) Then, hold a class discussion, asking students the following questions:

- What are some things in the video that surprised you?
- Have you ever been fooled by misleading statistics?
- What can we do to protect ourselves from being fooled by misleading statistics?
- What can we do to avoid using misleading statistics in our marketing campaign?

Discuss the following points with students based on McKee (2013):

- Find relevant and original sources
- Don't reword any statistics
- Always cite your sources
- Keep statistics up-to-date

- Double-check any data that sound too good to be true
- If you can, do your own research
- Don't lie or use false data

ELA Class: Display a short commercial script. These scripts are available online and can be found by an internet search using keywords such as "sample commercial scripts." "Voice Over Sample Scripts" at *www.voices.com/blog/voice-over-sample-scripts/* is one example. Read the script aloud, modeling the prescribed delivery, then call student attention to any casting, target, and delivery suggestions provided. Discuss how the target market may have affected marketer decisions about the tone and content of the ad.

Provide additional examples, and have students try out the delivery of the commercial messages following the suggestions provided within the scripts. Have students work in pairs to practice and deliver the commercial scripts, and have the rest of the class guess the target demographics.

Remind students that media messages are carefully constructed and that the person constructing the message has a goal in mind for that message. With commercials, the goal is to sell the product to a targeted market. Briefly discuss the five media literacy concepts listed on *www.medialit.org/reading-room/five-key-questions-form-foundation-media-inquiry*, guiding students to consider how each of the concepts plays out in the commercial texts they've been reading.

Give students copies of or access to information from the U.S. Small Business Administration (SBA) on how to design advertising for your product (found at *www.sba.gov/managing-business/running-business/marketing/advertising-basics*), explaining that this is an informational text produced to inform advertisers about how to construct an effective advertising campaign. Review strategies addressed so far for reading informational texts, including discussion of intended audience and structure of the article (and how that structure can help them comprehend the article).

Have students read and then discuss the article on how to design advertising for the assigned product. Highlight the steps the SBA suggests for creating an advertising campaign, including the following:

1. Identify the purpose of the advertising and how much money you have to spend.
2. Brainstorm about the features of your product you want to point out and identify your audience, and identify your competition.
3. Research your audience and your competition.

4. Decide where and how you will advertise (television, social media, newspapers, signs, or other means of advertising).
5. Create your advertising pieces, being sure to identify your product in all your advertising.

Working in groups, have students begin developing an advertising campaign for the product selected for the Product Design Challenge. The group members should collaborate to compose a plan (steps 1–3). Once this initial plan is articulated, meet with each group individually to help them determine what media to use and discuss the steps they need to take to design and complete their marketing product (steps 4 and 5). Provide students with ample time in class to work together, ready access to media tools (from technology to art supplies), and examples of similar work. The ideal classroom environment for this extended production stage of the lesson is a guided workshop, with the teacher acting as a facilitator, supporting idea development, providing feedback through the process, and helping as needed. Students will share their resulting advertisements in marketing presentations given at the end of the unit.

Show students examples of well-known logos to see how quickly they can identify the company and product. Explain that a logo is a constructed symbol that stands for a company's product. Engage students in exploring a variety of logos together. Discuss how marketers' decisions, such as use of color (e.g., bold and energizing or soft and soothing) and line (e.g., thick or fine, curvy or blocked), seem to fit with the product image they want to convey. Discuss why some particularly well-known companies might not use text in their logos. A well-designed logo will communicate and appeal to target consumers. Most effective logos include simple text that identifies the company, color that is eye-catching and appealing, and a simple graphic design that is easily recognizable and easy to reproduce.

Social Studies Connection: In this lesson, students explore social justice issues related to products and marketing. It is important for them to think about the sources of the products they produce, use, and purchase. Ask students the following questions: "How many of you love to eat chocolate? How is chocolate made? Where does cocoa come from?"

Tell students that they are going to watch a video that shows how the cocoa trade has affected one community in Ghana. Tie in geography here by asking students to find Ghana on a map, and tell them some important facts about Ghanaian culture, geography, and economy. If possible, use community resources and have someone from Ghana come in to share about his or her culture, geography, and economy.

Show students a video such as "History of Cocoa in Ghana" at *www.youtube.com/watch?v=oZ15kLPLVYY* or have them conduct research in teams about the history of cocoa production in Ghana. As students watch the video or conduct their research, have them write down some things that they notice, using the following prompts: What are

some issues associated with cocoa plantations? What is fair trade? How is life in Ghana different than life where you live?

Discuss the benefits of fair trade with students. Ask students to discuss as teams whether their module challenge products are fair trade products.

Elaboration/Application of Knowledge

Connection to the Challenge: On the last two days of the module, have students present their products, packaging designs, and campaigns to the panel of "company executives" and subject their product packaging to a crush test.

Science Connection: Students now apply what they have learned about media messages, statistics, and marketing to learn about the role media play in distributing ideas about science. Ask students the following questions:

- What role do media play in distributing ideas about science?
- How do social media play into this?
- What about the spreading of false science or misconceptions?

Tell students that each team should create a social media campaign that will spread the information they have learned about how we catch the common cold. Tell them to be sure that they are spreading correct information. Students should prepare their presentations to share the following day. Presentations need to include the following (see Rubric 9, p. 155):

- A description of the campaign's audience
- Factual information about ways to catch and prevent the common cold
- Graphics and images
- Concise and grammatically correct writing

Students should give their presentations to the class. You can then share their campaigns via social media if desired. Ask students, "How will you use this information to help you with your module project?"

Mathematics Class: Assign each group one of the following marketing claims or others you find in news reports to explore:

- Exaggerated health benefits of Activia yogurt and DanActive dairy drink
- Taco Bell's "seasoned beef" contents
- Olay's Definity eye cream's use of retouched photos
- Rice Krispies' claim to support immunity

- New Balance or Skechers shoes promise of stronger, shapelier legs
- Kashi's "all natural" label
- Eclipse gum's claim to kill germs

Have the teams research how the companies used misleading statistics to sell these products, and then share their findings with the class. What happened to the company? What could it have done differently? Then, explain to students that companies spend a lot of money on advertising. How do these companies decide whether the money is well spent? They use mathematics that we use in our classroom and daily lives to solve problems. Have students each work with a partner to find a solution to this problem that a typical company might experience: Over the past year, your company has spent $10,000 on social media bloggers to draw attention to the company. This campaign brought in 200 potential customers, with a 10% retention rate (meaning that the company now has 20 new customers). These customers had an average (mean) order of 500 items. If it costs your company $3 to make the product, and it sells each item for $10.50, how could you calculate the company's gross profit? Did your company make money after the marketing costs are factored in? What is the company's marketing return on investment percentage?

Define any the terms your students may need defined. You may wish to explain that marketers use measures known as *metrics* in their daily practice. See the key vocabulary in Table 4.8 (p. 130) for definitions of the following marketing metrics

- Return on investment (ROI)
- Gross profit
- Investment
- Revenue

Then, have students create their own generalizable model for determining the marketing return on investment (ROI) for the product described above. Have students share their solutions and models as a class. What does this ROI tell you for your company? Explain to students that this means that for every $1 spent on social media bloggers, it brought in $6.50 profit. Write a brief report to send to your company's CEO explaining your model and what the ROI was for the company.

Tell students that the CEO of the company responded to the report on the ROI with the following:

> Dear student, I am excited that our new investment in social media bloggers is bringing in so much revenue for our company. I would like for us to increase our social media blogger spending to $20,000 because it will double our return on investment.

Looking forward to our making more money in the next quarter.

Mark Jones, Company CEO

Assuming that doubling the number of bloggers will double the exposure, with a 10% customer retention rate, write a response to the CEO explaining why you do or do not support his proposal. Be sure to use the mathematical model you developed to present your evidence to the CEO. What would you recommend in order to double the ROI?

Students may struggle with calculating with large numbers and percentages. Some students may need direct instruction to help them solve the problems. However, you should provide direct instruction only after students have had ample time to productively struggle through the problem. Have students share their responses to the CEO with a partner, and then have them discuss their findings as a class.

STEM Research Notebook Prompt

Have students respond to the following question in their STEM Research Notebooks, and then discuss as a class: *How can marketers use mathematics to support company spending decisions?*

Students can apply what they learned about marketing ROI to calculate marketing expense to revenue. This is a metric used to show a company how much money it is spending on marketing compared with how much revenue it is generating. It is calculated by dividing the total marketing cost by the revenue generated. This gives the ratio of marketing cost to revenue. This cost includes salary for marketing employees. Ask students, How does this figure differ from ROI? You can give students a scenario similar to the social media marketing example above and have them solve this problem.

On the last few days of the lesson, have students spend mathematics class working on their module projects and presentations. In the final math class, have students answer the following questions in their STEM Research Notebooks:

- What have you learned about mathematics and its role in packaging, shipping, and marketing of products?
- What have you learned about misleading statistics and how to keep from getting duped as you observe ads critically?
- How have you used mathematics in your projects?
- How did you ensure that the information presented in your projects was mathematically accurate and appropriate?

Then, allow students to do a run-through of their presentations as time permits. Answer any questions students may have about the process.

ELA Class: Students should now develop effective logos for their products using the information they learned about logos in the previous class. Once their logos are finished, have students work on their advertising campaigns for the Product Design Challenge.

On the last two days of the module, have students present their products, packaging designs, and campaigns to the panel of "company executives" and subject their product packaging to a crush test.

After students have presented their products, packaging, and campaigns and have received feedback from the panel, student teams should discuss the feedback they received and brainstorm ways that their campaigns could be improved based upon the feedback.

Ask students, "What power does media literacy give you in your own life?" Discuss as a class. Then, have students work in their groups to create an action plan to address some of the issues they learned about in the module. Finally, ask each student group to create a public service message about buying fair trade cocoa and other fairly traded products.

STEM Research Notebook Prompt

Now that you have presented your products, packaging, and campaigns, reflect on your experience and answer the following questions:

- *How do you think your team's presentation went? What could have been improved about your presentation?*
- *What suggestions for improvement did the panel make?*
- *How can you improve your campaign based on those suggestions?*

Social Studies Connection: Have students work in pairs to research and draw the life cycle of their favorite chocolate bar, from seed to plate. Have them explore issues, both positive and negative, associated with the cocoa that is used to make their favorite chocolate bar. Have students share their life cycle and issue posters with the class. Tell them that some cocoa sources are responsibly produced. The following sources may help you share information with your students about different brands of chocolate:

- *www.thegoodtrade.com/features/fair-trade-chocolate*
- *www.slavefreechocolate.org/ethical-chocolate-companies*

Have students make an action plan to share how they could draw attention to some of the issues associated with cocoa. The following day, show students a video on interdependence in the world economy, such as the "Economic Interdependence Intro" at *www.youtube.com/watch?v=Li4TfR1xtZM*. Have students explore concepts such as economic environment, economy, economic interdependence, emerging markets, free trade, imports, and exports (see Table 4.8 on p. 130 for definitions). Have students choose a country and explore these concepts regarding their country.

The next day, have students present their findings to the class. Then, explain to students that much of the world's economy is interdependent. This is an abstract concept that may be difficult for students to understand, so you need to provide some concrete examples to students of how everything is related. Provide students with a modern case study, such as the economic downturn in Greece and how this has been having a global impact. Engage students in a discussion about ways that we can address issues associated with interdependence. Have students discuss in teams the impact that the U.S. economy has on other countries, then have teams share their ideas with the class.

Evaluation/Assessment

Performance Tasks

- Rubric 1: Product Design Challenge Rubric (see p. 85)
- Rubric 9: Social Media Campaign Rubric (see p. 155)

Other Measures

- STEM Research Notebook prompts

INTERNET RESOURCES

How one catches a cold

- *www.everydayhealth.com/cold-and-flu/colds-and-the-weather.aspx*
- *www.medicalnewstoday.com/articles/166606.php*
- *www.webmd.com/cold-and-flu/news/20150106/researchers-probe-why-colds-are-more-likely-in-winter*
- *www.pbs.org/wgbh/nova/next/body/scientists-finally-prove-cold-weather-makes-sick*

Spurious correlations

- *www.bbc.com/news/magazine-27537142*

The Federal Trade Commission and its protections

- *www.ftc.gov/about-ftc/what-we-do*
- *www.ftc.gov/news-events/media-resources/truth-advertising*

Cases of false advertising and deceptive practices

- *www.ftc.gov/news-events/press-releases/2010/12/dannon-agrees-drop-exaggerated-health-claims-activia-yogurt*

- *www.businessinsider.com/false-advertising-scandals-2011-9*
- *www.huffingtonpost.com/2011/01/25/taco-bell-beef-lawsuit_n_813185.html*
- *www.huffingtonpost.com/2014/05/16/companies-lied-to-you_n_5318940.html*
- *www.huffingtonpost.com/2012/08/29/new-balance-toning-shoe-settlement_n_1839537.html*

Misleading statistics

- *http://ed.ted.com/lessons/how-statistics-can-be-misleading-mark-liddell*
- *http://faculty.atu.edu/mfinan/2043/section31.pdf*
- *www.cracked.com/photoplasty_2052_19-ways-you-can-make-statistics-say-whatever-you-want*
- *www.truthpizza.org/logic/stats.htm*
- *www.econoclass.com/misleadingstats.html*

Marketing return on investment and other metrics

- *www.marketingmo.com/campaigns-execution/how-to-calculate-roi-return-on-investment*
- *www.investopedia.com/articles/personal-finance/053015/how-calculate-roi-marketing-campaign.asp*
- *http://blog.hubspot.com/marketing/math-behind-common-marketing-metrics*

Marketing and media literacy

- *www.cio.com/article/2377257/online-marketing/7-ways-to-create-a-successful-integrated-marketing-campaign.html*
- *www.entrepreneur.com/article/234832*
- *www.medialit.org*
- *www.sba.gov/managing-business/running-business/marketing/advertising-basics*

Information on Ghana and the cocoa trade

- *www.britannica.com/place/Ghana*
- *www.ghanaweb.com/GhanaHomePage/history*
- *www.infoplease.com/country/ghana.html*
- *http://crawfurd.dk/africa/ghana_timeline.htm*
- *www.ancient.eu/Ghana*

- *http://thegrio.com/2013/11/02/why-ghana-is-fast-becoming-a-hub-for-african-americans*
- *www.bbc.com/news/world-africa-34936103*
- *www.foodispower.org/slavery-chocolate*
- *http://equalexchange.coop/products/chocolate/faqs/what-makes-fair-trade-cocoa-and-chocolate-products-different*

Information on logos

- *http://diply.com/famous-logos-hidden-messages?publisher=what-facts&config=20*
- *www.smashingmagazine.com/2009/08/vital-tips-for-effective-logo-design*

Media literacy core concepts and key questions

- *www.medialit.org/reading-room/five-key-questions-form-foundation-media-inquiry*

"Voice Over Sample Scripts"

- *www.voices.com/blog/voice-over-sample-scripts*

"Pete's Dragon Passamaquaddy" video

- *www.youtube.com/watch?v=QHboMLW-Zn0*

"The Story of the Snake Oil Salesman" video

- *www.youtube.com/watch?v=titzrDTfp70*

"Does Being Cold Make You Sick?" video

- *www.youtube.com/watch?t=10&v=RWiOhlqEDz4*

"Flu Attack! How a Virus Invades Your Body" video

- *www.youtube.com/watch?t=206&v=Rpj0emEGShQ*

"Influenza Animation—Flu Virus Mechanism" video

- *www.youtube.com/watch?v=YSgkoldBNkI*

"Viruses: Virus Replication and the Mysterious Common Cold" video

- *www.youtube.com/watch?v=PHp6iYDi9ko*

"Airborne Commercial" video

- *www.youtube.com/watch?v=pDoG3gvE8Nc*

"Does Airborne Really Stave Off Colds?"

- *http://abcnews.go.com/GMA/OnCall/story?id=1664514&page=1*

"The Secret Behind Coca-Cola Marketing Strategy" video

- *www.youtube.com/watch?v=XhMVWzVXNNk*

"Don't Be Fooled by Bad Statistics" video

- *www.youtube.com/watch?v=jguYUbcIv8c*

Information on cocoa sources

- *www.thegoodtrade.com/features/fair-trade-chocolate*
- *www.slavefreechocolate.org/ethical-chocolate-companies*

"Economic Interdependence Intro" video

- *www.youtube.com/watch?v=Li4TfR1xtZM*

Rubric 9: Social Media Campaign Rubric

Name ______________________________

Criteria	Yes/No	Comments
A description of the audience was included and accurately represented the target market.		
Factual information about ways to catch and prevent the common cold was included in the presentation.		
Graphics and images were used to enhance the presentation.		
The presentation was written in a concise and grammatically correct manner.		

REFERENCES

Center for Media Literacy (CML). 2015. *Media literacy: A definition and more. www.medialit.org/media-literacy-definition-and-more.*

Dellwo, A. 2017. Double-blind, placebo-controlled clinical trial basics. *VeryWell,* December 7. *www.verywell.com/double-blind-placebo-controlled-clinical-trial-715861.*

Federal Trade Commission (FTC). 2008. Makers of Airborne settle FTC charges of deceptive advertising; agreement brings total settlement funds to $30 million. *www.ftc.gov/news-events/press-releases/2008/08/makers-airborne-settle-ftc-charges-deceptive-advertising.*

Gandhi, L. 2013. A history of "snake oil salesmen." *Code Switch: Word Watch,* National Public Radio, August 26. *www.npr.org/sections/codeswitch/2013/08/26/215761377/a-history-of-snake-oil-salesmen.*

LandLearn NSW. 2013. *What is sustainability? www.landlearnnsw.org.au/sustainability/what-is-sustainability.*

McKee, S. 2013. Sharing unfounded market statistics can hurt your credibility. *SurveyGizmo* (blog), September 18. *www.surveygizmo.com/survey-blog/marketers-sharing-unfounded-market-statistics-can-hurt-our-credibility.*

National Science Teachers Association (NSTA). 2000. NSTA position statement: The nature of science. *www.nsta.org/about/positions/natureofscience.aspx.*

Ohler, M., and P. Samuel. 2013. Five successful ideation session essentials. *Innovation Excellence* (blog), October 28. *www.innovationexcellence.com/blog/2013/10/28/five-successful-ideation-session-essentials.*

Sustainable Packaging Coalition (SPC). 2011. Definition of sustainable packaging. *https://sustainablepackaging.org/wp-content/uploads/2017/09/Definition-of-Sustainable-Packaging.pdf.*

WorldAtlas. 2017. Top 10 cocoa producing countries. April 25. *www.worldatlas.com/articles/top-10-cocoa-producing-countries.html.*

TRANSFORMING LEARNING WITH PACKAGING DESIGN AND THE *STEM ROAD MAP CURRICULUM SERIES*

Carla C. Johnson

This chapter serves as a conclusion to the Packaging Design integrated STEM curriculum module, but it is just the beginning of the transformation of your classroom that is possible through use of the *STEM Road Map Curriculum Series*. In this book, many key resources have been provided to make learning meaningful for your students through integration of science, technology, engineering, and mathematics, as well as social studies and English language arts, into powerful problem- and project-based instruction. First, the Packaging Design curriculum is grounded in the latest theory of learning for children in middle school specifically. Second, as your students work through this module, they engage in using the engineering design process (EDP) and build prototypes like engineers and STEM professionals in the real world. Third, students acquire important knowledge and skills grounded in national academic standards in mathematics, English language arts, science, and 21st century skills that will enable their learning to be deeper, retained longer, and applied throughout, illustrating the critical connections within and across disciplines. Finally, authentic formative assessments, including strategies for differentiation and addressing misconceptions, are embedded within the curriculum activities.

The Packaging Design curriculum in the Represented World STEM Road Map theme can be used in single-content middle school classrooms (e.g., science) where there is only one teacher or expanded to include multiple teachers and content areas across classrooms. Through the exploration of the Product Design Challenge, students engage in a real-world STEM problem on the first day of instruction and gather necessary knowledge and skills along the way in the context of solving the problem.

The other topics in the *STEM Road Map Curriculum Series* are designed in a similar manner, and NSTA Press has additional volumes in this series for this and other grade levels and plans to publish more. The volumes covering Innovation and Progress have been published and are as follows:

- Innovation and Progress
  - *Amusement Park of the Future, Grade 6*
  - *Construction Materials, Grade 11*
  - *Harnessing Solar Energy, Grade 4*
  - *Transportation in the Future, Grade 3*
  - *Wind Energy, Grade 5*

The tentative list of other books includes the following themes and subjects:

- The Represented World
  - Car crashes
  - Changes over time
  - Improving bridge design
  - Patterns and the plant world
  - Radioactivity
  - Rainwater analysis
  - Swing set makeover
- Cause and Effect
  - Influence of waves
  - Hazards and the changing environment
  - The role of physics in motion
- Sustainable Systems
  - Creating global bonds
  - Composting: Reduce, reuse, recycle
  - Hydropower efficiency
  - System interactions

- Optimizing the Human Experience
  - Genetically modified organisms
  - Mineral resources
  - Rebuilding the natural environment
  - Water conservation: Think global, act local

If you are interested in professional development opportunities focused on the STEM Road Map specifically or integrated STEM or STEM programs and schools overall, contact the lead editor of this project, Dr. Carla C. Johnson (*carlacjohnson@purdue.edu*), associate dean and professor of science education at Purdue University. Someone from the team will be in touch to design a program that will meet your individual, school, or district needs.

APPENDIX

CONTENT STANDARDS ADDRESSED IN THIS MODULE

NEXT GENERATION SCIENCE STANDARDS

Table A1 (p. 162) lists the science and engineering practices, disciplinary core ideas, and crosscutting concepts this module addresses. The supported performance expectations are as follows:

- MS-ETS1-1. Define the criteria and constraints of a design problem with sufficient precision to ensure a successful solution, taking into account relevant scientific principles and potential impacts on people and the natural environment that may limit possible solutions.
- MS-ETS1-2. Evaluate competing design solutions using a systematic process to determine how well they meet the criteria and constraints of the problem.
- MS-ETS1-3. Analyze data from tests to determine similarities and differences among several design solutions to identify the best characteristics of each that can be combined into a new solution to better meet the criteria for success.
- MS-PS1-3. Gather and make sense of information to describe that synthetic materials come from natural resources and impact society.
- MS-ESS3-3. Apply scientific principles to design a method for monitoring and minimizing a human impact on the environment.

Table A1. *Next Generation Science Standards (NGSS)*

Science and Engineering Practices
ASKING QUESTIONS AND DEFINING PROBLEMS Asking questions and defining problems in grades 6–8 builds on grades K–5 experiences and progresses to specifying relationships between variables, and clarifying arguments and models. • Define a design problem that can be solved through the development of an object, tool, process, or system and includes multiple criteria and constraints, including scientific knowledge that may limit possible solutions.
OBTAINING, EVALUATING, AND COMMUNICATING INFORMATION Obtaining, evaluating, and communicating information in 6–8 builds on K–5 and progresses to evaluating the merit and validity of ideas and methods. • Gather, read, and synthesize information from multiple appropriate sources and assess the credibility, accuracy, and possible bias of each publication and methods used, and describe how they are supported or not supported by evidence.
PLANNING AND CARRYING OUT INVESTIGATIONS Planning and carrying out investigations in 6–8 builds on K–5 experiences and progresses to include investigations that use multiple variables and provide evidence to support explanations or solutions. • Conduct an investigation to produce data to serve as the basis for evidence that meet the goals of an investigation.
ANALYZING AND INTERPRETING DATA Analyzing data in 6–8 builds on K–5 experiences and progresses to extending quantitative analysis to investigations, distinguishing between correlation and causation, and basic statistical techniques of data and error analysis. • Analyze and interpret data to determine similarities and differences in findings.
CONSTRUCTING EXPLANATIONS AND DESIGNING SOLUTIONS Constructing explanations and designing solutions in 6–8 builds on K–5 experiences and progresses to include constructing explanations and designing solutions supported by multiple sources of evidence consistent with scientific ideas, principles, and theories. • Apply scientific principles to design an object, tool, process, or system.
ENGAGING IN ARGUMENT FROM EVIDENCE Engaging in argument from evidence in 6–8 builds on K–5 experiences and progresses to constructing a convincing argument that supports or refutes claims for either explanations or solutions about the natural and designed world(s). • Use an oral and written argument supported by evidence to support or refute an explanation or a model for a phenomenon. • Evaluate competing design solutions based on jointly developed and agreed-upon design criteria.

Continued

Table A1. *(continued)*

Disciplinary Core Ideas
PS1.A. STRUCTURE AND PROPERTIES OF MATTER • Each pure substance has characteristic physical and chemical properties (for any bulk quantity under given conditions) that can be used to identify it.
PS1.B. CHEMICAL REACTIONS • Substances react chemically in characteristic ways. In a chemical process, the atoms that make up the original substances are regrouped into different molecules, and these new substances have different properties from those of the reactants.
ESS3.C. HUMAN IMPACTS ON EARTH SYSTEMS • Human activities have significantly altered the biosphere, sometimes damaging or destroying natural habitats and causing the extinction of other species. But changes to Earth's environments can have different impacts (negative and positive) for different living things.
ETS1.A. DEFINING AND DELIMITING ENGINEERING PROBLEMS • The more precisely a design task's criteria and constraints can be defined, the more likely it is that the designed solution will be successful. Specification of constraints includes consideration of scientific principles and other relevant knowledge that are likely to limit possible solutions.
ETS1.B. DEVELOPING POSSIBLE SOLUTIONS • There are systematic processes for evaluating solutions with respect to how well they meet the criteria and constraints of a problem. • Sometimes parts of different solutions can be combined to create a solution that is better than any of its predecessors.
ETS1.C. OPTIMIZING THE DESIGN SOLUTION • Although one design may not perform the best across all tests, identifying the characteristics of the design that performed the best in each test can provide useful information for the redesign process—that is, some of those characteristics may be incorporated into the new design.
Crosscutting Concepts
CAUSE AND EFFECT • Relationships can be classified as causal or correlational, and correlation does not necessarily imply causation.
STRUCTURE AND FUNCTION • Structures can be designed to serve particular functions by taking into account properties of different materials, and how materials can be shaped and used.

Continued

Table A1. *(continued)*

Crosscutting Concepts *(continued)*
INTERDEPENDENCE OF SCIENCE, ENGINEERING, AND TECHNOLOGY • Engineering advances have led to important discoveries in virtually every field of science, and scientific discoveries have led to the development of entire industries and engineered systems.
INFLUENCE OF SCIENCE, ENGINEERING, AND TECHNOLOGY ON SOCIETY AND THE NATURAL WORLD • All human activity draws on natural resources and has both short- and long-term consequences, positive as well as negative, for the health of people and the natural environment. • The uses of technologies and any limitations on their use are driven by individual or societal needs, desires, and values; by the findings of scientific research; and by differences in such factors as climate, natural resources, and economic conditions. Thus, technology use varies from region to region and over time.
SCALE, PROPORTION, AND QUANTITY • Phenomena that can be observed at one scale may not be observable at another scale.
SYSTEMS AND SYSTEM MODELS • Systems may interact with other systems; they may have sub-systems and be a part of larger complex systems.
SCIENCE IS A HUMAN ENDEAVOR • Scientists and engineers are guided by habits of mind such as intellectual honesty, tolerance of ambiguity, skepticism, and openness to new ideas.

Source: NGSS Lead States. 2013. *Next Generation Science Standards: For states, by states.* Washington, DC: National Academies Press. *www.nextgenscience.org/next-generation-science-standards.*

Table A2. Common Core Mathematics and English Language Arts (ELA) Standards

MATHEMATICAL PRACTICES AND CONTENT	READING AND WRITING STANDARDS
MATHEMATICAL PRACTICES • MP1. Make sense of problems and persevere in solving them. • MP3. Construct viable arguments and critique the reasoning of others. • MP4. Model with mathematics. • MP5. Use appropriate tools strategically. • MP6. Attend to precision. **MATHEMATICAL CONTENT** • 6.GA.2. Find the volume of a right rectangular prism with fractional edge lengths by packing it with unit cubes of the appropriate unit fraction edge lengths, and show that the volume is the same as would be found by multiplying the edge lengths of the prism. • 6.GA.4. Represent three-dimensional figures using nets made up of rectangles and triangles, and use the nets to find the surface area of these figures. Apply these techniques in the context of solving real-world and mathematical problems. • 6.SP.A.1. Recognize a statistical question as one that anticipates variability in the data related to the question and accounts for it in the answers. • 6.SP.A.3. Recognize that a measure of center for a numerical data set summarizes all of its values with a single number, while a measure of variation describes how its values vary with a single number. • 6.SP.B.5.B. Describing the nature of the attribute under investigation, including how it was measured and its units of measurement. • 6.SP.B.5.C. Giving quantitative measures of center (median and/or mean) and variability (interquartile range and/or mean absolute deviation), as well as describing any overall pattern and any striking deviations from the overall pattern with reference to the context in which the data were gathered.	**READING STANDARDS** • RH.6–8.4. Determine the meaning of words and phrases as they are used in a text, including vocabulary specific to domains related to history/social studies. • RI.6.2. Determine a central idea of a text and how it is conveyed through particular details; provide a summary of the text distinct from personal opinions or judgments. • RI.6.6. Determine an author's point of view or purpose in a text and explain how it is conveyed in the text. **WRITING STANDARDS** • W.6.4. Produce clear and coherent writing in which the development, organization, and style are appropriate to task, purpose, and audience. • W.6.6. Use technology, including the Internet, to produce and publish writing as well as to interact and collaborate with others; demonstrate sufficient command of keyboarding skills to type a minimum of three pages in a single sitting. • WHST.6–8.4. Produce clear and coherent writing in which the development, organization, and style are appropriate to task, purpose, and audience. • WHST.6–8.6. Use technology, including the Internet, to produce and publish writing and present the relationships between information and ideas clearly and efficiently. • WHST.6–8.8. Gather relevant information from multiple print and digital sources, using search terms effectively; assess the credibility and accuracy of each source; and quote or paraphrase the data and conclusions of others while avoiding plagiarism and following a standard format for citation. • WHST.6.10. Write routinely over extended time frames (time for reflection and revision) and shorter time frames (a single sitting or a day or two) for a range of discipline-specific tasks, purposes, and audiences.

Continued

Table A2. *(continued)*

MATHEMATICAL CONTENT *(continued)*	
• 6.EEA.2. Write, read, and evaluate expressions in which letters stand for numbers. • 6.EEB.6. Use variables to represent numbers and write expressions when solving a real-world or mathematical problem; understand that a variable can represent an unknown number, or, depending on the purpose at hand, any number in a specified set.	

Source: National Governors Association Center for Best Practices and Council of Chief State School Officers (NGAC and CCSSO). 2010. *Common core state standards.* Washington, DC: NGAC and CCSSO.

Table A3. 21st Century Skills From the Framework for 21st Century Learning

21st Century Skills	Learning Skills and Technology Tools	Teaching Strategies	Evidence of Success
INTERDISCIPLINARY THEMES	• Global Awareness • Financial, Economic, Business, and Entrepreneurial Literacy • Civic Literacy • Environmental Literacy • Health Literacy (understanding viruses and how we can prevent them from spreading)	• Teach students about other cultures and nations. • Have students consider how to make appropriate personal economic choices. • Teach students about the role of the economy in society. • Have students consider the global implications of civic decisions. • Teach students about the environment and circumstances and conditions affecting it. • Have students consider environmental issues and make conclusions about effective solutions. • Teach students about ways to avoid spreading colds through health and hygiene measures.	• Students understand how their purchasing decisions impact their fellow human beings in other nations. • Students have learned how they have been a target of marketers and understand the power of their purchasing decisions. • Students understand how their purchasing decisions at home have an impact on others. • Students understand the importance of purchasing sustainable packages and recycling. • Students are aware of how viruses can and cannot spread.

Continued

Table A3. *(continued)*

21st Century Skills	Learning Skills and Technology Tools	Teaching Strategies	Evidence of Success
LEARNING AND INNOVATION SKILLS	• Creativity and Innovation • Critical Thinking and Problem Solving • Communication and Collaboration	• Use a wide range of idea creation techniques. • Have students create new ideas and determine if they are worthwhile. • Have students elaborate, refine, and analyze their own ideas to improve and maximize creative efforts. • Ask students to use critical thinking to solve real-world problems. • Have students collaborate with others to solve problems. • Have students communicate skills in a variety of forms and contexts. • Model and discuss critically listening to decipher meaning of marketing messages.	• Students develop innovative solutions, product packaging, project logos, and marketing strategies. • Students evaluate marketing messages and were able to decipher their meaning. • Students show critical thinking and problem solving in addressing the module challenge • Students communicate their ideas and thoughts in a variety of ways (e.g., social media campaign, PowerPoint presentation, letter to a company, professional e-mail.) • Students present their marketing project to a panel of "company executives."
INFORMATION, MEDIA AND TECHNOLOGY SKILLS	• Information Literacy • Media Literacy • ICT Literacy	• Have students use nonfiction text to research sampling techniques and develop their mathematical models. • Help students use multimedia tools to present their findings.	• Students present to peers, teachers, and a panel of "company executives" using multimedia tools. • Students undersatnd the use of nonfiction text.

Continued

Table A3. *(continued)*

21st Century Skills	Learning Skills and Technology Tools	Teaching Strategies	Evidence of Success
LIFE AND CAREER SKILLS	• Flexibility and Adaptability • Initiative and Self-Direction • Social and Cross-Cultural Skills • Accountability, Leadership, and Responsibility	• Provide guidelines for effective peer critique and how to use this feedback to improve presentation. • Establish collaborative learning expectations. • Scaffold completion of tasks.	• Students provide feedback to peers on presentations and use feedback to improve presentations.

Source: Partnership for 21st Century Learning. 2015. Framework for 21st Century Learning. *www.p21.org/our-work/p21-framework.*

Table A4. English Language Development Standards

ELD STANDARD 1: SOCIAL AND INSTRUCTIONAL LANGUAGE
English language learners communicate for Social and Instructional purposes within the school setting.
ELD STANDARD 2: THE LANGUAGE OF LANGUAGE ARTS
English language learners communicate information, ideas and concepts necessary for academic success in the content area of Language Arts.
ELD STANDARD 3: THE LANGUAGE OF MATHEMATICS
English language learners communicate information, ideas and concepts necessary for academic success in the content area of Mathematics
ELD STANDARD 4: THE LANGUAGE OF SCIENCE
English language learners communicate information, ideas and concepts necessary for academic success in the content area of Science
ELD STANDARD 5: THE LANGUAGE OF SOCIAL STUDIES
English language learners communicate information, ideas and concepts necessary for academic success in the content area of Social Studies.

Source: WIDA. 2012. 2012 amplification of the English language development standards: Kindergarten–grade 12. *www.wida.us/standards/eld.aspx.*

INDEX

Page numbers printed in **boldface type** indicate tables, figures, or handouts.

P

T